OBJECTIONS SUSTAINED

Subversive Essays on Evolution, Law & Culture

PHILLIP E. JOHNSON

InterVarsity Press
Downers Grove, Illinois

InterVarsity Press
P.O. Box 1400, Downers Grove, IL 60515
World Wide Web: www.ivpress.com
E-mail: mail@ivpress.com

InterVarsity Press®️ is the book-publishing division of InterVarsity Christian Fellowship/USA®️, a student movement active on campus at hundreds of universities, colleges and schools of nursing in the United States of America, and a member movement of the International Fellowship of Evangelical Students. For information about local and regional activities, write Public Relations Dept., InterVarsity Christian Fellowship/USA, 6400 Schroeder Rd., P.O. Box 7895, Madison, WI 53707-7895.

Cover illustration: Kurt Mitchell

ISBN 0-8308-1941-X

Printed in the United States of America ♺

Library of Congress Cataloging-in-Publication Data

Johnson, Phillip E., 1940-
    Objections sustained : subversive essays on evolution, law, and
culture / Phillip E. Johnson.
      p.   cm.
    Includes bibliographical references.
    ISBN 0-8308-1941-X (alk. paper)
    1. Science—Social aspects. 2. Science—Philosophy. 3. Evolution
(Biology)—Philosophy. I. Title.
Q175.5.J6 1998
303.48'3—dc21
                                                   98-6866
                                                  CIP

19	18	17	16	15	14	13	12	11	10	9	8	7	6	5	4	3	2	1
13	12	11	10	09	08	07	06	05	04	03	02	01	00	99	98			

To the members of the Wedge,
present and future

Introduction

My professional life was transformed in the academic year 1987-1988, which I spent on sabbatical leave from Berkeley at University College, London. With no teaching duties and no definite plan for research, I was ready for a new direction. It came when I encountered the literature of Darwinism, and I quickly became fascinated with the ways in which Richard Dawkins, Stephen Jay Gould and other prominent Darwinists argued for their theory—or perhaps I should say for their differing theories that all fit somewhere under the umbrella formed by that elastic term *evolution*.

In *Darwin on Trial* (IVP, 1993) I explained why Darwinian theory is scientifically on very weak ground. The theory is sustained largely by a propaganda campaign that relies on all the usual tricks of rhetorical persuasion: hidden assumptions, question-begging statements of what is at issue, terms that are vaguely defined and change their meaning in midargument, attacks on straw men, selective citation of evidence, and so on. The theory is also protected by its cultural importance. It is the officially sanctioned creation story of modern society, and publicly funded educational authorities spare no effort to persuade the public to believe in it.

The ways the campaign is conducted and some of the people who have taken leading roles in it are the subject of the nine essays in part one of this book. We begin with Thomas Henry Huxley, Darwin's first and greatest disciple, and end with the current controversy over whether "evolution" as taught in the schools should be explicitly atheistic or merely implicitly naturalistic. There have always been bitter divisions in the Darwinian camp, but in the end the rivals have banded together to fight the common enemy, believers in supernatural creation. These essays explain the rivalries and divisions. I hope they will prepare readers to understand the growing crisis of Darwinism, which I expect to emerge into public view sometime in the first decade of the twenty-first century.

As the official creation story of the intellectuals who dominate public policy and education, evolutionary naturalism has enormous importance for the culture in general. This was the subject of my second book, *Reason in the Balance* (IVP, 1995), and it is the starting point for the thirteen book-review essays in part two. I find the book review a thoroughly satisfying form. I review only books that I find interesting, whether I agree with the author or not, and the discipline of writing a review motivates me to concentrate on (and appreciate) the author's thought much more intensely than I otherwise would. Authors want to be reviewed, and they are usually grateful if the reviewer demonstrates that he has read the book carefully, no matter what opinions he expresses about it. For this reason authors tend to appreciate my reviews, and I have made some marvelous new friends this way.

Most of the essays in both sections appeared in one or the other of two estimable journals: *First Things* and *Books & Culture*. I am very grateful to the editors for the support they have given me, and I want to take this opportunity to urge you to consider subscribing to at least one or the other—and preferably both. You will be amply rewarded by the quality of the writing you will receive. *First Things* is the world's leading journal for writing from all perspectives about the fascinating issues of religion and culture. *Books & Culture,* under the umbrella of *Christianity Today*, is taking mighty steps toward achieving that long-awaited goal, the renaissance of the evangelical mind.

If you see any merit in the essays that follow, please consider supporting the journals that have supported my writing.*

*For subscriptions to *First Things*, telephone 1-800-783-4903. For subscriptions to *Books & Culture*, telephone 1-800-523-7964.

Part 1

Nine Essays About Darwinists & Darwinism

1

The Pope of the New Religion

THIS FIRST ESSAY IN THE VOLUME, A REVIEW OF ADRIAN DESMOND'S Huxley: From Devil's Disciple to Evolution's High Priest *(Addison-Wesley, 1997), was actually the last to be published, running in early 1998 in* The Washington Times. *I place it at the beginning because it deals with the early triumph of Darwinism, when Thomas Huxley won the nineteenth-century culture war for the professional intellectuals against the church and the aristocracy. Desmond is also the author, with James Moore, of an excellent biography of Charles Darwin:* Darwin: The Life of a Tormented Evolutionist *(Warner, 1992). Another biographical work I recommend, especially for Christians, is James Moore's* The Darwin Legend *(Baker, 1994). Moore debunks the legend, still persistent among Christians, that Darwin repudiated his theory and accepted Christ on his deathbed. On the contrary, Darwin was never more than a nominal Christian, and he had become a confirmed agnostic by the time of his death. Darwin avoided any public display of his religious unbelief because he did not want to shock his devout wife or to prejudice the public against his theory.*

A drian Desmond is an esteemed intellectual historian of the "sociology of knowledge" school, which emphasizes the role of scientific ideas in cultural and political conflicts. No individual invites study from this perspective more than Thomas Henry Huxley, who used Darwin's theory as a battering ram to champion the cause of professional scientists against the aristocratic patronage system, and thus to make "what you know" more important than "who you know."

In the process, Huxley effectively disestablished the Church of England and replaced it with a new priesthood whose creed was evolutionary naturalism. Desmond's subtitle and chapter headings indicate the primacy of Huxley's role as religious revolutionary. Huxley went from being the Devil's Disciple (early champion of Darwin's heresy) to Evolution's High Priest (when Darwinism itself became the orthodoxy). Desmond introduces his subject at the beginning as "The Apostle Paul of the New Teaching" and leaves him at the end as "Pope Huxley," whose agnosticism (paradoxically) was "the last act of the Protestant Reformation."

One whole section is titled "The New Luther," and the comparison is instructive. Luther undermined the power of the medieval clergy by encouraging the common people to read the Bible for themselves. Huxley lectured on science to packed audiences of working people, who saw that they could use the new knowledge to challenge the power of the aristocracy and the industrialists. Both revolutionaries were displeased by the radical consequences that some derived from their teaching. Luther took the side of the nobility when the peasants revolted, and Huxley used his authority as pope of Darwinism to defend competition and meritocracy against radical socialists.

The crowning irony is that Huxley was a Darwinist only in the broadest sense of the term. He believed in evolutionary naturalism and in the exclusive authority of scientists to answer questions like how we came to be created, but he had little regard for Darwin's mechanism of natural selection except as a weapon to fend off the clergy. According to Desmond (p. 392), "Where Darwin had nature select from tiny variations, Huxley had been happy with larger jumps; where Darwin transported his species to islands by wind and raft, Huxley had invoked drowned continents. At times all they seemed to share was a faith in evolutionary naturalism."

The continuing split between the Darwinists and the Huxleyists, who are united only in their mutual hatred for creationism, came to the surface again in 1997 when the Huxleyist Stephen Jay Gould launched a furious attack on "Darwinian Fundamentalists" (especially Richard Dawkins and Daniel Dennett) in *The New York Review of Books*. Both contending parties made it clear that what unites the warring factions of evolutionists is faith in evolutionary naturalism and opposition to the possibility of divine creation. As Gould's ally Richard Lewontin wrote, also in the course of disparaging Dawkins, "we cannot allow a Divine Foot in the door."* That very factor explains, however, why Darwin's name and not Huxley's is practically synonymous with *evolution,* and why Dawkins can trump Gould's arguments whenever he has to. Huxleyism lacks a credible mechanism for explaining the immense complexity of even the simplest organisms, and the complexity-building mechanism is essential for shutting out the persistent Divine Foot.

The interaction between Huxley's scientific rationalism and Adrian Desmond's sociology of knowledge illustrates the most profound conflict now going on in the university world—the conflict over who studies whom. Scientific materialists want to explain everything, including the mind and its ideas, as the products of physical interactions. Sociologists of knowledge want to explain the cultural forces, and financial incentives, that cause ideologies like scientific materialism to prosper. As Desmond explains it, "temporal benefits were the prize as a sectarian science contrasted itself with the wealthy Church.

---

*The Lewontin and Gould essays cited in this paragraph are discussed in essays 7 and 8 ("The Unraveling of Scientific Materialism" and "The Gorbachev of Darwinism") in this book.

... The new science was deliberately made to tell against the clergy's supernatural sanction" (p. 435).

Being able to assign an irrational cause for the other person's beliefs is a powerful weapon in a cultural conflict. As I drafted this review, I read in my morning paper of a neuroscience conference where brain researchers announced that they had discovered a "God module," meaning a part of the brain that affects the intensity of religious belief. "It is not clear why such dedicated machinery for religion may have evolved," the researchers reported. You may be sure that scientific materialists will never discover a "materialist module," meaning a brain part that causes people to fantasize that they can explain the mind in strictly materialist terms. Sociologists of knowledge do something like that, however, when they explain the very material incentives that can lead scientists to get carried away with their methodology.

Desmond's book will tell you a lot about Thomas Huxley, probably more than you really want to know. This meticulous, detailed history is well worth the effort, however, because it carries an important lesson for the present time. In 1981 the U.S. National Academy of Sciences resolved that "religion and science are separate and mutually exclusive realms of human thought whose presentation in the same context leads to misunderstanding of both scientific theory and religious belief." Probably some of the Academy members who voted for that resolution were so naive as to believe it, while others cynically saw it as a useful stick with which to beat the creationists. The life of Thomas Huxley is the best answer to such nonsense. In reality scientists (like other people) are obsessed with the God question, and the whole point of evolutionary naturalism is to keep that Divine Foot, and the people gathered behind it, from getting inside the door.

2

What Is Darwinism?

ORIGINALLY DELIVERED IN 1993 AS A LECTURE AT A SYMPOSIUM AT Hillsdale College, the following essay predictably drew the wrath of theistic evolutionists. In fact, a furious biology professor in the audience shouted out, "Don't give up your day job, professor!" The lecture was then published by the Hillsdale College Press in the collection Man and Creation: Perspectives on Science and Theology, *edited by Michael Bauman (1993). I particularly recommend this essay for beginning students of the creation-evolution conflict, because it explains the all-important Darwinian vocabulary. As long as Darwinists control the definitions of the key terms, their system is unbeatable regardless of the evidence.*

There is a popular television game show called *Jeopardy!* in which the usual order of things is reversed. Instead of being asked a question to which they must supply the answer, contestants are given the answer and asked to provide the appropriate question. This format suggests an insight that is applicable to law, to science and indeed to just about everything. The important thing is not necessarily to know all the answers but rather to know what question is being asked.

That insight is the starting point for my inquiry into Darwinian evolution and its relationship to creation, because Darwinism is the answer to two very different kinds of questions. First, Darwinian theory tells us how a certain amount of diversity in life forms can develop once we have various types of complex living organisms already in existence. If a small population of birds happens to migrate to an isolated island, for example, a combination of inbreeding, mutation and natural selection may cause this isolated population to develop different characteristics from those possessed by the ancestral population on the mainland. When the theory is understood in this limited sense, Darwinian evolution is uncontroversial and has no important philosophical or theological implications.

But evolutionary biologists are not content merely to explain how variation occurs within limits. They aspire to answer a much broader question: how complex organisms like birds and flowers and human beings came into existence in the first place. The Darwinian answer to this second question is that the creative force that produced complex plants and animals from single-celled predecessors over long stretches of geological time is essentially the same as the mechanism that produces variations in flowers, insects and domestic animals before our very eyes. In the words of Ernst Mayr, the dean of living Darwinists, "transspecific evolution [i.e., macroevolution] is nothing but an extrapolation and magnification of the events that take place within populations and species."

Neo-Darwinian evolution in this broad sense is a philosophical doctrine so lacking in empirical support that Mayr's successor at Harvard, Stephen Jay Gould, once pronounced it in a reckless moment to be "effectively dead." Yet neo-Darwinism is far from dead; on the contrary, it is continually proclaimed in the textbooks and the media as unchallengeable fact. How does it happen that so many scientists and intellectuals, who pride themselves on their empiricism and open-mindedness, continue to accept an unempirical theory as scientific fact? The answer to that question lies in the definition of five key terms. The terms are *creationism, evolution, science, religion* and *truth*. Once we understand how these words are used in evolutionary discourse, the continued ascendancy of neo-Darwinism will be no mystery and we need no longer be deceived by claims that the theory is supported by "overwhelming evidence."

I should warn at the outset, however, that using words clearly is not the innocent and peaceful activity most of us may have thought it to be. There are powerful vested interests in this area which can thrive only in the midst of ambiguity and confusion. Those who insist on defining terms precisely and using them consistently may find themselves regarded with suspicion and hostility and even accused of being enemies of science. But let us accept that risk and proceed to the definitions.

Creation and Creationism
The first word is *creationism*, which literally means no more than the belief that there is a Creator. In Darwinist usage, which dominates not only the popular and professional scientific literature but also the media, a creationist is a person who takes the creation account in the book of Genesis to be true in a very literal sense. The earth was created in a single week of six twenty-four-hour days no more than ten thousand years ago; the major features of the earth's geology were produced by Noah's flood; and there have been no major innovations in the forms of life since the beginning. It is a major theme of Darwinist propaganda that the only persons who have any doubts about Darwinism are young-earth creationists of this sort, who are always portrayed as rejecting the clear and convincing evidence of science to preserve a religious prejudice. The implication is that citizens of modern society are faced with a choice that is really no choice at all. Either they reject science altogether and retreat to a premodern worldview,

or they believe everything the Darwinists tell them.

In a broader sense, however, a creationist is simply a person who believes in the existence of a Creator who brought about the existence of the world and its living inhabitants in furtherance of a purpose. Whether the process of creation took a single week or billions of years is relatively unimportant from a philosophical or theological standpoint. Creation by gradual processes over geological ages may create problems for biblical interpretation, but it creates none for the basic principle of theistic religion. And creation in this broad sense, according to a 1991 Gallup poll, is the creed of 87 percent of Americans. If God brought about our existence for a purpose, then the most important kind of knowledge to have is knowledge of God and of what he intends for us. Is creation in that broad sense consistent with evolution?

Evolution and Naturalism

The answer is absolutely not, when evolution is understood in the Darwinian sense. To Darwinists evolution means *naturalistic* evolution, because they insist that science must assume that the cosmos is a closed system of material causes and effects which can never be influenced by anything outside of material nature—by God, for example. In the beginning, an explosion of matter created the cosmos, and undirected, naturalistic evolution produced everything that followed. From this philosophical standpoint it follows deductively that from the beginning no intelligent purpose guided evolution. If intelligence exists today, that is only because it has itself evolved through purposeless material processes.

A materialistic theory of evolution must inherently invoke two kinds of processes. At bottom the theory must be based on chance, because that is what is left when we have ruled out everything involving intelligence or purpose. Theories that invoke *only* chance, though, are unbelievable. One thing everyone acknowledges is that living organisms are enormously complex—far more so than, say, a computer or an airplane. That such complex entities came into existence simply by chance is clearly less credible than that they were designed and constructed by a Creator. To back up their claim that this appearance of intelligent design is an illusion, Darwinists need to provide some complexity-building force that is mindless and purposeless. Natural selection is by far the most plausible candidate.

If we assume that random genetic mutations provided the new genetic information needed, say, to give a small mammal a start toward wings, and if we assume that each tiny step in the process of wing building gave the animal an increased chance of survival, then natural selection ensured that the favored creatures would thrive and reproduce. It follows as a matter of logic that wings can and will appear as if by the plan of a designer. Of course if wings or other improvements do *not* appear, the theory explains their absence just as well. The needed mutations didn't arrive, or "developmental constraints" closed off certain possibilities, or natural selection favored something else. There is no requirement that any of this speculation be confirmed by either experimental or fossil evidence. For Darwinists, just being able to imagine the process is sufficient to confirm that something like that must have happened.

Dawkins calls the process of creation by mutation and selection "the blind watchmaker," by which label he means that a purposeless, materialistic designing force substitutes for the "watchmaker" deity of natural theology. The creative power of the blind watchmaker is supported by only very slight evidence, such as the famous example of a moth population in which the percentage of dark moths increased during a period when birds were better able to see light moths against smoke-darkened background trees. This may be taken to show that natural selection can do *something,* but not that it can create anything that was not already in existence. Even such slight evidence is more than sufficient, however, because evidence is not really necessary to prove something that is practically self-evident. The existence of a potent blind watchmaker follows deductively from the philosophical premise that nature had to do its own creating. There can be argument about the details, but if God was not in the picture something very much like Darwinism simply has to be true, regardless of the evidence.

Science and Its Paradigms

That brings me to my third term, *science*. We have already seen that Darwinists assume as a matter of first principle that the history of the cosmos and its life forms is fully explicable on naturalistic principles. This reflects a philosophical doctrine called scientific naturalism, which is said to be a necessary consequence of the inherent limitations of science. What scientific naturalism does, however, is to transform the limitations of science into limitations on reality, in the interest of maximizing the explanatory power of science and its practitioners. It is, of course, entirely possible to study organisms scientifically on the premise that they were all created by God, just as scientists study airplanes and even works of art without denying that these objects are intelligently designed. The problem with allowing God a role in the history of life is not that science would cease but rather that scientists would have to acknowledge the existence of something important that is outside the boundaries of natural science. For scientists who want to be able to explain everything—and "theories of everything" are now openly anticipated in the scientific literature—this is an intolerable possibility.

The second feature of scientific naturalism that is important for our purpose is its set of rules governing the criticism and replacement of a paradigm. A paradigm is a general theory, like the Darwinian theory of evolution, that has achieved general acceptance in the scientific community. The paradigm unifies the various specialties that make up the research community and guides research in all of them. Thus zoologists, botanists, geneticists, molecular biologists and paleontologists all see their research as aiming to fill out the details of the Darwinian paradigm. If molecular biologists see a pattern of apparently neutral mutations that have no apparent effect on an organism's fitness, they must find a way to reconcile their findings with the paradigm's requirement that natural selection guides evolution. This they can do by postulating a sufficient number of invisible adaptive mutations, which are deemed to be accumulated by natural selection. Similarly, if paleontologists see new fossil species appearing suddenly in the fossil record and remaining basically unchanged thereafter, they must perform whatever contortions are necessary to force this recalcitrant evidence into a model of incremental change through the accumulation of micromutations.

Supporting the paradigm may even require what in other contexts would be called deception. As Niles Eldredge has candidly admitted, "We paleontologists have said that the history of life supports [the story of gradual adaptive change], all the while knowing it does not." Eldredge explained that this pattern of misrepresentation occurred because of "the certainty so characteristic of evolutionary ranks since the late 1940s, the utter assurance not only that natural selection operates in nature, but that we know precisely how it works." This certainty produced a degree of dogmatism that Eldredge says resulted in the relegation to the "lunatic fringe" of paleontologists who reported that "they saw something out of kilter between contemporary evolutionary theory, on the one hand, and patterns of change in the fossil record on the other."* Under the circumstances, prudent paleontologists understandably swallowed their doubts and supported the ruling ideology. To abandon the paradigm would be to abandon the scientific community; to ignore the paradigm and just gather the facts would be to earn the demeaning label "stamp collector."

As many philosophers and scientists have observed, the research community does not abandon a paradigm in the absence of a suitable replacement. This means that negative criticism of Darwinism, however devastating it may appear to be, is essentially irrelevant to the professional researchers. The critic may point out, for example, that the evidence that natural selection has any creative power is somewhere between weak and nonexistent. That is perfectly true, but to

---

*See Niles Eldredge, *Time Frames* (Simon & Schuster, 1985), pp. 93, 144.

Darwinists the more important question is this: If natural selection did not do the creating, what did? "God" is obviously unacceptable, because such a being is unknown to science. "We don't know" is equally unacceptable, because to admit ignorance would be to leave science adrift without a guiding principle. To put the problem in the most practical terms: it is impossible to write or evaluate a grant proposal without a generally accepted theoretical framework.

The paradigm rule explains why Gould's acknowledgment that neo-Darwinism is "effectively dead" had no significant effect on the Darwinist faithful, or even on Gould himself. Gould made that statement in a paper predicting the emergence of a new general theory of evolution, one based on the macromutational speculations of the Berkeley geneticist Richard Goldschmidt.* When the new theory did not arrive as anticipated, the alternatives were either to stick with Ernst Mayr's version of neo-Darwinism or to concede that biologists do not after all know of a naturalistic mechanism that can produce biological complexity. That was no choice at all. Gould had to beat a hasty retreat back to classical Darwinism to avoid giving aid and comfort to the enemies of scientific naturalism, including those disgusting creationists.

Having to defend a dead theory tooth and nail can hardly be a satisfying activity, and it is no wonder that Gould lashes out with fury at one such as I, who calls attention to his predicament.** I do not mean to ridicule Gould, however, because I have a genuinely high regard for the man as one of the few Darwinists who have recognized the major problems with the theory and reported them honestly. His tragedy is that he cannot admit the clear implications of his own thought without effectively resigning from science.

---

*Stephen Jay Gould, "Is a New and General Theory of Evolution Emerging?" *Paleobiology* 6 (1980): 119-30, reprinted in the collection *Evolution Now: A Century After Darwin,* ed. Maynard Smith (Freeman, 1982).

**See Stephen Jay Gould's attack on my book *Darwin on Trial,* "Impeaching a Self-Appointed Judge," *Scientific American,* July 1992, pp. 118-22. My detailed response is contained in the epilogue to the revised edition of *Darwin on Trial* (InterVarsity Press, 1993).

The continuing survival of Darwinist orthodoxy illustrates Thomas Kuhn's remark that the accumulation of anomalies never in itself falsifies a paradigm, because "to reject one paradigm without substituting another is to reject science itself." This practice may be appropriate as a way of carrying on the professional enterprise called science, but it can be grossly misleading when it is imposed on persons who are asking questions other than the ones scientific naturalists want to ask. Suppose, for example, that I want to know whether God really had something to do with creating living organisms. A typical Darwinian response is that there is no reason to invoke supernatural action, because Darwinian selection was capable of performing the job. To evaluate that response, I need to know whether natural selection really has the fantastic creative power attributed to it. It is not a sufficient answer to say that scientists have nothing better to offer. The fact that scientists don't like to say "We don't know" tells me nothing about what they really *do* know.

I am not suggesting that scientists have to change their rules about retaining and discarding paradigms. All I want them to do is to be candid about the disconfirming evidence and admit, if it is the case, that they are hanging on to Darwinism only because they prefer a shaky theory to no theory at all. What they insist on doing, however, is to present Darwinian evolution to the public as a fact that every rational person is expected to accept. If there are reasonable grounds to doubt the theory, such dogmatism is ridiculous, whether or not the doubters have a better theory to propose.

Religion and Reason

To believers in creation, the Darwinists seem thoroughly intolerant and dogmatic when they insist that their own philosophy must have a monopoly in the schools and the media. The Darwinists do not see themselves that way, of course. On the contrary, they often feel aggrieved when creationists (in either the broad or the narrow sense) ask to have their own arguments heard in public and fairly considered.

To insist that schoolchildren be taught that Darwinian evolution is a fact is in their minds merely to protect the integrity of science education; to present the other side of the case would be to allow fanatics to force their opinions on others. Even college professors have been forbidden to express their doubts about Darwinian evolution in the classroom, and it seems to be widely believed that the Constitution not only permits but actually requires such restrictions on academic freedom. To explain this bizarre situation, we must define our fourth term: *religion*.

Suppose a skeptic argues that evidence for biological creation by natural selection is obviously lacking and that in the circumstances we ought to give serious consideration to the possibility that the development of life required some input from a preexisting, purposeful Creator. To scientific naturalists this suggestion is "creationist" and therefore unacceptable in principle, because it invokes an entity unknown to science. What is worse, it suggests the possibility that this Creator may have communicated in some way with humans. In that case there could be real prophets—persons with a genuine knowledge of God who are neither frauds nor dreamers. Such persons could conceivably be dangerous rivals to the scientists as cultural authorities.

The strategy that naturalistic philosophy has worked out to prevent this problem from arising is to label naturalism as science and theism as religion. The former is then classified as *knowledge* and the latter as mere *belief.* The distinction is of critical importance, because only knowledge can be objectively valid for everyone; belief is valid only for the believer and should never be passed off as knowledge. The student who thinks that 2 and 2 make 5, or that water is not made up of hydrogen and oxygen, or that the theory of evolution is not true, is not expressing a minority viewpoint. He or she is ignorant, and the job of education is to cure that ignorance and replace it with knowledge. Students in the public schools are thus to be taught at an early age that "evolution is a fact," and as time goes by they will gradually learn that evolution means naturalism.

In short, the proposition that God was in any way involved in our creation is effectively outlawed and implicitly negated. This is because naturalistic evolution is by definition in the category of scientific knowledge. What contradicts knowledge is implicitly false or imaginary. That is why it is possible for scientific naturalists in good faith to claim on the one hand that their science says nothing about God and to claim on the other hand that they have said everything that can be said about God. In naturalistic philosophy both propositions are at bottom the same. All that needs to be said about God is that there is nothing to be said of God, because on that subject we can have no knowledge.

Truth

Our fifth and final term is *truth*. Truth as such is not a particularly important concept in naturalistic philosophy. The reason for this is that "truth" suggests an unchanging absolute, whereas scientific knowledge is a dynamic concept. What was knowledge in the past is not knowledge today, and the knowledge of the future will surely be far superior to what we have now. Only naturalism itself and the unique validity of science as the path to knowledge are absolutes. There can be no criterion for truth outside of scientific knowledge, no mind of God to which we have access.

This way of understanding things persists even when scientific naturalists employ religious-sounding language. For example, the physicist Stephen Hawking ended his famous book *A Brief History of Time* with the prediction that human beings might one day "know the mind of God." This phrasing cause some friends of mine to form the mistaken impression that he had some attraction to theistic religion. In context Hawking was not referring to a supernatural eternal being, however, but to the possibility that scientific knowledge will eventually become complete and all-encompassing because it will have explained the movements of material particles in all circumstances.

The monopoly of science in the realm of knowledge explains why evolutionary biologists do not find it meaningful to address the question whether the Darwinian theory is true. They will gladly concede that the theory is incomplete and that further research into the mechanisms of evolution is needed. At any given point, though, the reigning theory of naturalistic evolution represents the state of scientific knowledge about how we came into existence. Scientific knowledge is by definition the closest approximation to absolute truth available to us. To ask whether this knowledge is true is therefore to miss the point and to betray a misunderstanding of "how science works."

So far I have described the metaphysical categories by which scientific naturalists have excluded the topic of God from rational discussion and thus ensured that Darwinism's fully naturalistic creation story is effectively true by definition. There is no need to explain why atheists find this system of thought-control congenial. What is a little more difficult to understand, at least at first, is the strong support Darwinism continues to receive in the Christian academic world. Attempts to investigate the credibility of the Darwinist evolution story are regarded with little enthusiasm by many leading Christian professors of science and philosophy, even at institutions that are generally regarded as conservative in theology.

Given that Darwinism is inherently naturalistic and therefore antagonistic to the idea that God had anything to do with the history of life, and that it plays the central role in ensuring agnostic domination of the intellectual culture, one might have supposed that Christian intellectuals (along with religious Jews) would be eager to find its weak spots. Instead the prevailing view among Christian professors has been that Darwinism—or "evolution," as they tend to call it—is unbeatable and that it can be interpreted to be consistent with Christian belief.

Theistic Evolution

In fact Darwinism *is* unbeatable as long as one accepts the thought categories of scientific naturalism that I have been describing. The problem is that those same thought categories relegate Christian theism, or any other theism, to the never-never land of subjective belief. If science has exclusive authority to tell us how life was created, and if science is committed to naturalism, and if science never discards a paradigm until it is presented with an acceptable naturalistic alternative, then Darwinism's position is impregnable within science. The same reasoning that makes Darwinism inevitable, however, also bans God from taking any action within the history of the cosmos, which means that it makes theism illusory. Theistic naturalism is self-contradictory.

Some hope to avoid the contradiction by asserting that naturalism rules only within the realm of science and that there is a separate realm called "religion" in which theism can flourish. The problem with this arrangement, as we have already seen, is that in a naturalistic culture scientific conclusions are considered to be knowledge, or even fact. What is outside of fact is fantasy, or at best subjective belief. Theists who accommodate scientific naturalism therefore may never affirm that their God is *real* in the same sense that evolution is real.

Under naturalistic rules, evolution is confirmed by scientific evidence and hence is real for all reasonable purposes; whatever role God may have taken in the process is inherently superfluous and invisible—except to believers. This distinction is essential to the entire mindset that produced Darwinism in the first place. If God exists, he could certainly work through mutation and selection if that is what he wants to do, but he could also create by some means totally outside the ken of our science.

Once we put God into the picture, there is no good reason to attribute the creation of biological complexity to random mutation and natural selection. Direct evidence that these mechanisms have substantial creative power is not to be found in nature, the laboratory or the fossil record. An essential step in the reasoning that establishes that Darwinian selection created the wonders of biology, therefore, is that nothing else was available. Theism is by definition the doctrine that something else *was* available. Allow a preexisting supernatural intelligence to guide evolution, and this omnipotent being can do a whole lot more than that.

Of course theists can think of evolution as God-guided whether naturalistic Darwinists like it or not. The trouble with having a private definition for theists, however, is that the scientific naturalists have the power to decide what *evolution* means in public discourse, including science classes in the public schools. If theistic evolutionists broadcast the message that evolution as *they* understand it is harmless to theistic religion, they are misleading their constituents—unless they add a clear warning that the version of evolution advocated by the entire body of mainstream science is something else altogether. That warning is never clearly delivered, because the main point of theistic evolution is to preserve peace with the mainstream scientific community. Theistic evolutionists therefore unwittingly serve the purposes of scientific naturalists, by helping persuade the religious community to lower its guard against the incursion of naturalism.

We are now in a position to answer the question with which I began: What is Darwinism? Darwinism is a theory of empirical science only at the level of microevolution, where it provides a framework for explaining such things as the diversity that arises when small populations become reproductively isolated from the main body of their species. As a general theory of biological creation, Darwinism is not empirical at all. Rather, it is a necessary implication of a philosophical doctrine called scientific naturalism, which is based on the a priori assumption that God was always absent from the realm of nature. As such, evolution in the Darwinian sense is inherently antithetical to theism, although evolution in some entirely different and nonnaturalistic sense could conceivably have been God's chosen method of creation.

What Is Darwinism?

In 1874 the great Presbyterian theologian Charles Hodge asked the question "What is Darwinism?" After a careful and thoroughly fairminded evaluation of the doctrine, his answer was unequivocal: "It is

Atheism." Another way to state the proposition is to say that Darwinism is the answer to a specific question that grows out of philosophical naturalism. To return to the game of Jeopardy with which we started, let us say that Darwinism is the answer. What, then, is the question? The question is: How must creation have occurred if we assume that God had nothing to do with it?

Theistic evolutionists accomplish very little by trying to Christianize the answer to a question that comes straight out of the agenda of scientific naturalism. Instead we need to challenge the assumption that the only questions worth asking are the ones that assume that naturalism is true.

3

Domesticating
Darwin

SECULARISTS LIKE THOMAS HUXLEY WELCOMED THE CONCEPT OF
*naturalistic evolution as a means of liberating humankind from the
tyranny of priests and religious dogmatists. Yet Darwin's mechanism
of natural selection threatened to substitute a new tyranny for the old
one. If human qualities and behavior patterns have been fashioned by
natural selection, it may be difficult or even impossible for us to
change them. The existence of inequalities between races, social
classes and the sexes may be "natural," just as it is natural for many
social animals to establish hierarchies of dominance and submission.
For this reason many Darwinists of the left are vehemently hostile
toward any attempt to explain human qualities of behavior in reference
to natural selection, whether this is done under the old name of social
Darwinism or within modern reformulations such as sociobiology or
evolutionary psychology.*

*Contemporary Darwinists sometimes give the impression that so-
cial Darwinism was a later perversion of Darwin's biological theory*

rather than an integral part of it. In fact social Darwinism came first, and the biological theory grew out of the social speculations of such thinkers as Herbert Spencer and Thomas Malthus. Darwin wrote in his personal journal that the basic idea of natural selection came to him after he had read the work of Malthus, who had argued that welfare measures for the poor are futile because their beneficial effects are inevitably swamped by population growth. Starvation of the many leads to survival of the fittest and hence benefits the species in the long run. Perhaps we should not speak of "social Darwinism" but rather of "biological Spencerism." Social Darwinism later became an embarrassment, and in In Search of Human Nature: The Decline and Revival of Darwinism in American Social Thought *(Oxford University Press, 1991), the noted intellectual historian Carl Degler explains how it was banished to the outer margins of the intellectual world. My essay was originally published in the May 1993 issue of* First Things.

In the beginning, Charles Darwin explained how human beings evolved from animals by natural selection. One might have expected that this revelation would have a profound effect on disciplines like psychology and anthropology and that the newly established link to animals would become the foundation for the scientific study of human behavior. Carl Degler tells us that at first the Darwinian agenda did indeed dominate the human sciences. Darwin himself filled three chapters of *The Descent of Man* with arguments that the mental and moral faculties of human beings were derived from similar features of animals.

This explanatory project carried some extremely racist implications, however. Because Darwin was determined to establish human continuity with animals, he frequently wrote of "savages and lower races" as intermediate between animals and civilized people. Thus Degler observes that it was as much Darwin himself as any of the so-called social Darwinists who set the evolutionary approach to human behavior on a politically unacceptable course. "Thanks to

Darwin's acceptance of the idea of hierarchy among human societies,"
he tells us, "the spread and endurance of a racist form of social
Darwinism owes more to Charles Darwin than to Herbert Spencer."

A scientific grounding for racism is not the only unsavory heritage
of nineteenth-century Darwinism. Degler also cites Darwin's theories
about the intellectual inferiority of women and describes how Darwin's cousin Francis Galton employed Darwinian logic to advocate
an ambitious eugenics program to improve the breed. It is not surprising that such doctrines were unpopular with influential social scientists like the Columbia University anthropologist Franz Boas—and his
famous pupils Margaret Mead and Ruth Benedict—who were dedicated to egalitarianism and cultural relativism.

Degler rightly emphasizes, however, that the conflict of social
science with Darwin involved something much more fundamental
than the use of Darwinism to support specific social prejudices. A
foundational aspect of Darwinian biological theory conflicted with
optimistic views about how the human condition might be improved.
As Darwinism became neo-Darwinism, it became increasingly clear
that the theory ruled out the Lamarckian notion of the heritability of
acquired characteristics. At first reform-minded social scientists had
thought "evolution" would allow a better social environment to lead
to improvements in races or groups—improvements that would be
made permanent through inheritance. These hopes were dashed, however, when August Weissman cut off the tails of many generations of
mice, thus proving to the satisfaction of biologists that changes
affecting an animal's body or behavior during its lifetime are not
passed on to its offspring. Sigmund Freud, a law unto himself, continued to embrace the theory that acquired characteristics are heritable,
but theorists of lesser stature were constrained by the facts. As Degler
puts it, they had to choose "between biology—which no longer could
be seen as experientially cumulative—and culture, which was."

Naturally, the reformers chose the alternative that fitted their social
vision. Degler makes clear that the choice was primarily based on

ideological rather than scientific considerations. "The main impetus [for the shift from biology to culture] came from a wish to establish a social order in which innate and immutable forces of biology played no role in accounting for the behavior of social groups. . . . To the proponents of culture the goal was the elimination of nativity, race, and sex, and any other biologically based characteristic that might serve as an obstacle to an individual's self-realization."

Degler serves up some delightful quotes to illustrate how extreme the rejection of biology was—for example, Berkeley anthropologist Alfred Kroeber's 1915 declaration that "heredity cannot be allowed to have acted any part in history." The main rhetorical weapon employed on behalf of culture was to place an insurmountable burden of proof on anyone offering a biological explanation for some human traits or practice. "As Boas characteristically phrased it, if a biological explanation could not be conclusively proved, then culture must be the causal element." Behaviorism and egalitarianism were in, eugenics and intelligence testing were out. As the epitome of the transformation, Degler cites Gunnar Myrdal's immensely influential study of race relations in the United States, An American Dilemma. Myrdal wrote at the end of this book, "We have today in social science a greater trust in the improvability of man and society than we have ever had since the Enlightenment."

But this optimism was purchased at the cost of a certain intellectual incoherence. Degler considers it profoundly ironic that social scientists became convinced

> that human beings in their social behavior, alone among animals, have succeeded in escaping biology. . . . For that belief is accompanied by another deeply held conviction: that human beings, like all other living things, are products of the evolution that Charles Darwin explained with his theory of natural selection. The irony is almost palpable inasmuch as Darwin entertained no doubt that behavior was as integral a part of human evolution as bodily shape.

Rejection of Darwinism in human behavior furthered the professional independence of social science from biological science at the same time that it encouraged visions of human perfectibility. On the other hand, social scientists have always wanted to be recognized as genuine scientists, an aspiration requiring that they somehow connect their discipline to the rest of the scientific enterprise. The problem they faced after consolidating the triumph of culture was to find a way to reincorporate a certain amount of biology without allowing it to set the social science agenda or to impose limits on human possibilities. The obvious solution was to readmit biological theories of human behavior up to a point but to keep them on a short political leash.

The second half of Degler's book describes a "return of biology" in recent decades that has featured figures such as Edward O. Wilson, Konrad Lorenz, Nicolaus Tinbergen and William Hamilton. Degler insists that this renewed interest in biology

> is not simply a revival of repudiated ideas, like racism, sexism, or eugenics. The evolutionary approach to social science has no more place for them than has the currently dominant cultural interpretation. . . . Rather, the true aim of these social scientists who advocate a "return" is to ask how human beings fit into that which Darwin laid down over a century ago and which very few social scientists consciously repudiate—except when the behavior of human beings is included in it.

In short, biologically minded social scientists know better than to draw the politically incorrect inferences from their theory that led to the banishment of their predecessors. If there are any exceptions, they are tactfully ignored. Degler gives no consideration to the merits of theories that there may be inherited group differences in intelligence, for example. The possible existence of biology-based behavioral differences between men and women is discussed primarily from a feminist perspective, or at least by researchers respectful of feminist sensitivities. No one proposes a revival of the eugenics movement or even feels the need to explain why a eugenics program is not a good idea from a Darwinian standpoint.

The political leash is so securely fashioned that it hardly ever needs to be jerked. Nonetheless, some voices of the left like Harvard biologists Richard Lewontin and Stephen Jay Gould remain intensely suspicious of sociobiology even when it comes with a liberal spin. Their objection is not to the immediate political agenda of the sociobiologists but to the inherent potential of biological theories of human behavior to provide support for those who think the existing distribution of wealth and power is "natural." As a result, biological theorizing still occupies only a marginal position in the culture of academic social science.

Degler is always interesting and informative, but it is hard to take seriously a "return to biology" in social science that is so severely constrained by ideology. All that seems to have happened is the spreading of a Darwinian gloss on the ideals of the priesthood of culture. What Degler calls "social science" is less a science than a secularized version of liberation theology. This secular theology can tolerate a biology that is obedient to ideology, but it dares not risk the kind of uncontrolled scientific investigation that might tell us something about human nature and capabilities that we would rather not know.

Darwinism was valuable to anthropologists like Boas, Benedict, Kroeber and Mead because it did so much to discredit religious authority and thus seemed to leave humans free to chart their own course in a world without restrictions. Cultural anthropologists, psychologists and social engineers had to accept Darwinism to explain the history of life up to a point when human culture began, because they regarded theistic religion as so much superstition. But once Darwinism had served to liberate humanity from God, its continuing presence inherently threatened to legitimate restrictions on human possibilities by subjecting human behavior to the all-conquering power of natural selection. The reformers could tolerate Darwinism only if it respected the sacred doctrines of human equality and perfectibility and paid its way by helping the visionaries to claim the authority of science. If evolutionary biology has indeed returned to social science, it is by way of the servants' entrance.

4

Extinction
Bad Genes or Bad Luck?

DARWINISTS SPEAK OF "NATURAL SELECTION" AS IF IT WERE A CREATIVE force, but it is really just a fancy name for nonrandom death. From a Darwinian viewpoint, however, the presence of a lot of fossils of extinct creatures testifies to the emergence of newer, fitter forms of life that prevailed in the struggle for existence. Extinction of the unfit is also cited as evidence against intelligent design in biology. Why would a wise Creator design creatures that were unfit to survive? In fact, there is no reason to believe that extinct forms of life were any less fit to survive under the normal conditions of their time than are modern plants and animals, including ourselves. Whatever the true explanation of extinctions may be, the Darwinian explanation finds no support whatever in the fossil evidence, as David Raup explains in Extinction: Bad Genes or Bad Luck? *(Norton, 1991).*

This essay has a poignant history for me. While Darwin on Trial *was in page proofs, an editor from* The Atlantic *phoned with an offer to let me write a long article based on the book for his magazine. Alas,*

I had already published just such an article in First Things—*an estimable journal, but one of far less circulation. Under the circumstances the editor withdrew the offer but, as a consolation prize, allowed me to review the book of my choice. I picked the Raup volume and set to work establishing my thesis that the Darwinian theory of extinction cannot be separated from the Darwinian theory of biological creation.*

The piece barely made it to publication, as it was assigned to a subeditor who was an enthusiastic Darwinist and thought my line of reasoning was crazy. It was carried in the February 1992 issue. Letters to the editor were carried in three subsequent issues of the magazine. All the published letters were vehemently hostile, but Raup himself wrote to me privately and said I was right on target.

David Raup, a paleontologist at the University of Chicago, is a leading figure among scientists who have rehabilitated catastrophe as a respectable concept in the study of the earth's history. His book is a readable but state-of-the-art account of what scientists know about why dinosaurs and other fossil species aren't with us anymore. By concluding that species become extinct through bad luck rather than because they are unfit, Raup inadvertently raises some awkward questions about how those species got there in the first place.

Scientific thinking about extinctions has strayed far from the Darwinian principles that still define orthodoxy in the life sciences as a whole. According to Raup, the Darwinian theory that extinctions result from the slow and steady effects of biological competition is "appealing, and has been learned by generations of biology students." Nonetheless, "its verification from actual field data is negligible." Raup goes on to say, without really arguing the point, that abandonment of the Darwinian explanation of extinctions does not discredit Darwin's theory that today's species evolved from earlier species through natural selection. To explain what lies behind that disclaimer—and why it can't be accepted at face value—I need to explain something about the place of extinctions in Darwinian theory.

Darwinian evolution is best known as a theory about the origin of species, but it is equally a theory about how species become extinct. The most influential interpreter of fossil history before the triumph of Darwinism was the great French sage Georges Cuvier, today forgotten in the English-speaking world but in his time renowned as the Aristotle of biology. Cuvier reported that there had apparently been profound catastrophes in the earth's history. Entire groups of creatures disappeared abruptly from the rocks and just as abruptly were replaced by new groups, which repopulated the earth.

Because catastrophic extinctions and sudden creations are outside the normal range of human experience, they seem elusive to scientific investigation. It would be much more convenient for science if important changes in the earth's history resulted gradually from the uniform operation over immense time periods of natural forces that we can observe in operation today. The motto of uniformitarian science is "The present is the key to the past." It is a triumphant motto, because with such a key in its possession, science can in principle unlock all the secrets of the history of the earth. Scientists thus had a strong professional motivation to reinterpret the geological evidence according to rules that disallowed both catastrophes and sudden creations as scientific explanations.

The lawyer and geologist Charles Lyell established a rigorous uniformitarianism as the basis of geology, and Darwin extended Lyell's logic to biology. Species neither appeared nor disappeared suddenly, according to Darwin's classic *On the Origin of Species*. Rather, they evolved step by tiny step from earlier forms, owing to the accumulation of tiny favorable variations through natural selection. Species declined to extinction equally gradually, as they were supplanted by modified descendants or by competing species that were more proficient at surviving and reproducing. If Darwin was right, both the origin and the extinction of living forms occurred through the slow and steady action of forces—reproduction, inheritance and competition—that we see operating in everyday life.

Creation by natural selection and extinction by natural selection are not two separate processes but two aspects of the same process. In Darwinian terms, superior fitness means superior ability at leaving descendants. If evolution has furthered the development of capabilities like strength, vision and intelligence, it is only because organisms possessing these (inheritable) qualities have consistently left more descendants than have competing organisms that lack them. The more fit crowd out the less fit by definition, and there is no such thing as natural selection unless they do. In Darwin's words,

> The theory of natural selection is grounded on the belief that each new variety, and ultimately each new species, is produced and maintained by having some advantage over those with whom it comes into competition; and the consequent extinction of less favored forms almost inevitably follows.

Darwin cited this logical relationship between evolution and extinction to refute one of the most formidable objections to his theory. "Evolution" implies a continuous process of change, but nature is organized in discrete groups that seem isolated from one another. How could the same organism be the ancestor of both an insect and a vertebrate except by miraculous transformation? Where in the living world are the intermediate links that ought to exist if continuous change occurred? The answer, Darwin wrote, is that

> as natural selection acts solely by the preservation of profitable modifications, each new form will tend in a fully stocked country to take the place of, and finally to exterminate, its own less-improved parent or other less-favoured forms with which it comes into competition. Thus extinction and natural selection will, as we have seen, go hand in hand. Hence, if we look at each species as descended from some other unknown form, both the parent and all the transitional varieties will generally have been exterminated by the very process of formation and perfection of the new form.

If extinction and natural selection did not go hand in hand, some of the parent forms ought to have survived to the present—but none are to be found. It is therefore an essential element of Darwinism that species continually became extinct because they were less fit than their descendants or other rivals. And because superior fitness itself emerges very gradually, extinction of a competing species should also proceed gradually. What is true of individual species should be still more true of groups of species—families, orders, classes and so on.

Because of this logic, Darwin insisted that Cuvier's theory of periodic catastrophes had been thoroughly discredited and that, on the contrary, "there is reason to believe that the complete extinction of the species of a group is generally a slower process than their production." According to Darwin, the struggle for existence is so finely tuned that "the merest trifle would often give the victory to one organic being over another." He continued, "Nevertheless so profound is our ignorance, and so high our presumption, that we marvel when we hear of the extinction of an organic being: and as we do not see the cause we involve cataclysms to desolate the world." In short, Darwin maintained that earlier geologists had attributed extinctions to catastrophes not because that was a reasonable interpretation of the fossil evidence but because they were ignorant of the higher law of natural selection.

If one wanted to subject Darwin's theory to empirical testing, one way to do it would be to examine the history of extinctions. Does the evidence confirm that biological competition was frequently the cause of extinctions? Can the occurrence of a Darwinian extinction—by competition from a closely related rival—be confirmed in even a single case? To put the point the other way around: Have the paleontologists, despite their best efforts to see fossil history in a Darwinian light, found that Cuvier was much closer to the truth after all? The answers couldn't be clearer.

The story starts with the famous "K-T" (Cretaceous-Tertiary) extinction of sixty-five million years ago. The K-T is not the biggest of the Big Five mass extinctions that mark the close of various geological ages, but it is the most recent, and the most fascinating from an anthropocentric point of view. It eliminated the dinosaurs, the ammonites and a lot else, clearing the way for mammals to dominate the planet. The geologist Walter Alvarez (with his famous physicist father, Luis Alvarez, and others) startled the world in 1979 by attributing this mass extinction to the impact of a meteor or a comet, which caused a worldwide environmental disaster that disrupted the food chain. As Raup puts it, the first reaction of paleontologists schooled in Lyellian and Darwinian concepts was horror and disbelief: "It was like suggesting that the dinosaurs had been shot by little green men from a spaceship." Hardly more than a decade later, so much confirming evidence has piled up that Raup predicts that soon "it will be difficult to find anyone who ever doubted the impact-extinction link." In science as in war, victory has a thousand fathers; defeat is an orphan.

Was the K-T extinction (or were the Big Five as a group) an exception to a general Darwinian pattern of gradual extinctions by competition? Raup thinks that environmental disasters triggered by large meteor impacts may have caused most extinctions other than those caused by human beings. There is some evidence of a connection between major extinctions and known meteor impacts, but Raup concedes that the evidence is far from conclusive. What inclines him to the meteor theory is primarily the difficulty of attributing extinctions to more mundane causes.

It turns out to be very difficult to kill off a numerous and wide-ranging species unless a catastrophic "first strike" has severely depleted its numbers and restricted its range. Even the spectacular environmental stress induced by the last great ice age produced a relatively modest number of extinctions, although the casualties included such popular favorites as mammoths and saber-toothed tigers. It seems that something far outside the ordinary run of environmental hazards is needed to kill enough individuals to threaten a widely distributed species with extinction, and meteor impacts may be the least unlikely of the known alternatives.

Maybe most extinctions were triggered by giant meteor impacts and maybe they weren't. If it is difficult to determine what did cause the bulk of extinctions, it is much easier to determine what didn't. There is no hard evidence that any observable extinctions were caused by competition from closely related species. Raup notes that evolutionary biologists long emphasized competition as a cause of extinctions because the explanation "seemed self-evident," but when they actually tried to test the effect of competition, the results were negative. The only reason to attribute extinctions to Darwinian competition remains the theory itself.

Some groups survived mass extinctions, but this does not mean they were more fit in any Darwinian sense than the groups that died out. Characteristics that would aid survival under normal circumstances would not necessarily be of any use under the extreme conditions of a catastrophe. For example, the only thing that identifies mammals as more fit than dinosaurs is that some mammals happened to survive the K-T extinction. When a prominent Darwinist attributed the survival of some groups in a mass extinction to their possessing "resistance to extinction," his statement added nothing to the bare fact that they had survived.

But then what becomes of Darwinism? Raup answers that attributing extinctions to bad luck rather than bad genes does not discredit Darwin's theory of evolution by natural selection, because

> natural selection remains the only viable, naturalistic explanation we have for sophisticated adaptations like eyes and wings. We would not be here without natural selection. Extinction by bad luck merely adds another element to the evolutionary process, operating at the level of species, families, and classes, rather than the level of local breeding populations of single species. (p. 192)

The trouble with that disclaimer is that in Darwin's theory survival of the fittest and extinction of the less fit are the same thing, not two separable processes. When Darwinists say that natural selection preserves favorable mutations, they mean it selectively kills off organisms not possessing the favorable mutation. To allow mutation to make new eyes and wings step by step, natural selection has to be constantly eliminating parent organisms without eyes and wings, or with inferior eyes and wings. Darwinists have always considered extinct fossil species to be evidence that natural selection operates as the theory requires. If that is not the case, then where is the evidence that natural selection has ever done more than produce relatively trivial variations in local breeding populations? A natural selection that only creates and never destroys is a logical impossibility, because it wouldn't be doing any selecting.

There is a way out of this logical impasse, but it is one I think Raup or any other empirical scientist would be reluctant to take. Only a minority of species are fossilized, and hence only a minority of extinctions are recorded. Perhaps only the visible extinctions went according to Cuvier and the invisible ones were Darwinian. But withdrawing a subject from empirical investigation in order to protect a theory from falsification is hardly the scientific thing to do. Raup says that the study of extinctions was long neglected; it could be that the influence of Darwin kept it off limits.

5

The Storyteller
& the Scientist

THE FOLLOWING ESSAY REVIEWS MICHAEL BEHE'S DARWIN'S BLACK
Box: The Biochemical Challenge to Evolution *(Free Press, 1996) and
Richard Dawkins's* Climbing Mount Improbable *(Viking, 1996). It
was originally published in* First Things, *October 1996, and was later
published in translation in the Russian scientific journal* Poisk. *Here
I turn to the positive side of the Darwinian mutation/selection mechanism. Daniel Dennett (see "Daniel Dennett's Dangerous Idea" in this
book) has called Darwin's mechanism "the best idea anyone has ever
had." I would call it a brilliant hypothesis for its time, but one that
subsequent investigation has discredited. To show why, I compare
Dawkins's storytelling with the science of Michael Behe.*

Richard Dawkins began *The Blind Watchmaker,* his influential
restatement of Darwinism, with the observation that "biology
is the study of complicated things that give the appearance of
having been designed for a purpose." May we consider the possibility
that living organisms give that appearance because they actually *were*

designed? Dawkins, who is virtually the defining example of an uncompromising scientific materialist, meets that suggestion with the scorn he thinks it deserves. The point of evolutionary science, he says, is to explain how complex things get made from a simple start. An unevolved Designer who is presumably more complex than the things he designs just doesn't fit into that picture. In *Climbing Mount Improbable* Dawkins calls organisms "designoids"—meaning things that look exactly as if they were designed but must actually have been crafted by the "blind watchmaker"—that is, the mindless Darwinian forces of mutation and selection.

Biochemist Michael Behe answers that the blind watchmaker thesis is a relic of a nineteenth-century science that lacked the understanding of biological mechanisms that recent advances in molecular biology have provided. The biologists who established the still-dominant Darwinian orthodoxy thought of the cell as an undifferentiated blob of "protoplasm." Like a child imagining he might construct an airplane out of cardboard boxes and pieces of wood, they could blithely propose materialist evolutionary scenarios for biological systems because they had no idea of how those systems actually work. The organism (and especially the cell) was to them a "black box"—a machine that does wonderful things by some mechanism nobody knows.

Behe explains that biochemists are now able to explore part of the insides of that black box, and what they find inside is "irreducible complexity." A system is irreducibly complex if it is "composed of several well-matched, interacting parts that contribute to the basic function, wherein the removal of any one of the parts causes the system to effectively cease functioning." Life at the molecular level is replete with such systems, and biochemists do not even attempt to explain how any one of them could have come into existence by the Darwinian mechanism. The result of biochemical investigation of cellular mechanisms, according to Behe, "is a loud, clear, piercing cry of 'Design!' "

The Behe argument is as revolutionary for our time as Darwin's argument was for his. If true, it presages not just a change in a scientific theory but an overthrow of the worldview that has dominated intellectual life ever since the triumph of Darwinism, the metaphysical doctrine of scientific materialism or naturalism.

A lot is at stake, and not just in science. But can a fair scientific test be devised to judge the competing merits of the positions staked out by Dawkins and Behe? Not if the Designer is ruled out a priori by a philosophical dogma, but Dawkins has said that his position is falsifiable:

> One hundred and twenty-five years [after the publication of Charles Darwin's masterpiece *On the Origin of Species*], we know a lot more about animals and plants than Darwin did, and still not a single case is known to me of a complex organ that could not have been formed by numerous successive slight modifications. I do not believe that such a case will ever be found. If it is—it'll have to be a *really* complex organ, and . . . you have to be sophisticated about what you mean by 'slight'—I shall cease to believe in Darwinism." (*Blind Watchmaker,* p. 91)

Dawkins agrees that even a single irrefutable case of irreducible complexity would be fatal to Darwinism. Behe argues that there are many cases of irreducible complexity to be found at the molecular level, with more being discovered as the science progresses. What is more, he argues that the existence of irreducible complexity is *implicitly* accepted by the entire worldwide community of molecular biologists.

I emphasize that word *implicitly,* because most prominent molecular biologists definitely would not concede the point *explicitly.* Molecular biology is dominated by metaphysical materialists, many of whom will proclaim to every journalist in sight that their discipline confirms Darwinism in every detail. What molecular biology has to say is determined not by what the biologists say to a popular audience, however, or even to each other in conversation, but by what they publish in leading scientific journals. Behe reports that what they do not *ever* publish in those journals is detailed scenarios of how even a single complex molecular system could have evolved by a Darwinian process.

In short, the irreducible complexity of molecular systems is controversial among molecular biologists when it is presented as an *idea* with philosophical consequences, and tacitly accepted as unpleasant reality when it remains in the world of innocent fact. When some tactless individual does raise the possibility of design at a biological gathering, perhaps after a few drinks, the auditors greet the remark with the embarrassed silence that might follow the disclosure of some shameful secret. To understand why Behe's argument is so uncontested in the realm of *fact* and yet why so many scientists find the *concept* of irreducible complexity not only difficult to accept but even impossible to consider, we should start by summarizing the modern understanding of Darwinism, as set out by Richard Dawkins.

Is There a Staircase Up Mount Improbable?

Everybody agrees that organisms are extremely complex. As Dawkins puts it with his usual rhetorical skill:

> Physics books may be complicated, but . . . the objects and phenomena that a physics book describes are simpler than a single cell in the body of its author. And the author consists of trillions of those cells, many of them different from each other, organized with intricate architecture and precision-engineering into a working machine capable of writing a book. . . . Each nucleus . . . contains a digitally coded database larger, in information content, that all 30 volumes of the *Encyclopedia Britannica* put together. And this figure is for *each* cell, not all the cells of the body put together. (*Blind Watchmaker,* pp. 2-3)

That informational complexity is the summit of the "Mount Improbable" of his title. The living world contains innumerable such mountains of complexity, and the Darwinist must show how they can all be reached without the aid of a miraculous leap or a boost from some preexisting intelligence. Just as a mountain climber cannot jump to the top of the Matterhorn, a (relatively) simple organism like a bacterium cannot even conceivably become a complex plant or animal except in very gradual stages. Fossil experts like Stephen Jay Gould sometimes distinguish between "evolution" and "gradualism," primarily because they are trying to square the former with the fact that the fossil record does not reflect a pattern of gradual transformations, but evolution has to be gradual when it is employed to explain how an unintelligent process assembled all that complex genetic information.

If the blind watchmaker thesis is true, there must be a gradually ascending staircase from the base all the way to the summit. To restate the metaphor in biological language, there must have existed a continuous series of viable intermediate forms between the first replicating organism (whose origin is another subject) all the way to every complex type of organ system and organism that has ever existed. Each step upward in complexity has to be at least slightly fitter (at leaving descendants) than its predecessor, and the gap between the steps must be no wider than can be bridged by random mutation. On the whole that means *tiny* mutations because, according to Dawkins, mutations large enough to have visible effects are nearly always harmful. The gradual steps have to be virtually omnipresent; a few plausible sections of staircase here and there up the face of the mountain are not enough. As Dawkins concedes, even a single unclimbable precipice spoils the theory—although the difficulty in proving that any one precipice is truly unclimbable means that a great many examples will have to be considered.

Because of Dawkins's philosophical starting point (science goes from simple to complex), he does not regard the staircase as something whose existence needs to be proved, but rather as a logical necessity that needs only to be illustrated. The illustrations consist primarily of imaginative stories and computer simulations. Here, for example, is a synopsis of the Dawkins theory on the evolution of flight:

To begin with, an ancestor like an ordinary squirrel, living up trees without any special gliding membrane, leaps across short gaps. [It could leap farther if it had something to slow a fall.] So natural selection favors individuals with slightly pouchy skin around the arm or leg joints, and this becomes the norm. . . . Now any individuals with an even larger skin web can leap a few inches further. So in later generations this extension of skin becomes the norm, and so on. . . . It is easy to imagine true flapping flight evolving from repetition of the muscular movements used to control glide direction, so average time to landing is gradually postponed over evolutionary time.

Some biologists, however, prefer to see long-distance downhill gliding as the dead end of the tree-jumping line of evolution. True flight, they think, began on the ground rather than up trees. . . . There are some mammals such as kangaroos that propel themselves very fast on two legs, leaving their arms free to evolve in other directions. . . . But bipedal mammals don't seem to have taken the next step and evolved the power of flight. The only true flying mammals are bats, and their wing membrane incorporates the back legs as well as the arms. . . . Perhaps birds began flying by leaping off the ground, while bats began by gliding out of trees. Or perhaps birds too began by gliding out of trees. The debate continues. (*Climbing Mount Improbable*, pp. 108-13)

Many biologists call this kind of explanation a "Just-So Story" because it belongs to the realm of children's literature, not science. Dawkins is like the little boy who thought he could make an airplane by adding something that looks like a pair of wings to something that looks like a fuselage. How do you make a bat? No problem, boys and girls, and no need to consider the complications of biochemistry, physiology and development. Just wait for a squirrel population to grow wings, which it might do one way or another.

Dawkins's computer simulations of evolution have even less connection to biological reality. A computer program can be *designed* (the word deserves emphasis here) to do just about anything, including to mutate stick figures that look vaguely like animals (or trees) into all kinds of shapes. The eminent Darwinist John Maynard Smith dismissed the much more sophisticated computer simulations of Stuart Kauffmann as "fact-free science," because they have no connection to real biological mechanisms.

The Irreducible Complexity of Biochemistry

To move from Dawkins to Behe is like moving from the children's library to the laboratory. Do you want to know how vision might have evolved? Because the biochemistry of vision is a black box to Dawkins, he can speculate without impediment. There are well over forty different types of eyes, which, because of their fundamentally differing structure, must have evolved (whatever that means) separately. Some of these eyes are much simpler than others. All an evolutionary storyteller has to do is start with the apparently simplest version, ignore the neural equipment that has to be present for an organism to make any use of a "photon receptor," and spin a charming tale about how a tiny, primitive light-sensing cell might grow up to be a full-fledged eye. That's what Charles Darwin did in 1859, and Dawkins repackages the same story.

Behe gives us just a bare start toward understanding what a biochemically informed evolutionary theory has to explain:

> When light first strikes the retina a photon interacts with a molecule called 11-cis retinal, which rearranges within picoseconds to *trans*-retinal. (A picosecond is about the time it takes light to travel the breadth of a single human hair.) The change in the shape of the retinal molecule forces a change in the shape of the protein, rhodopsin, to which the retinal is tightly bound. The protein's metamorphosis alters its behavior. Now called metarhodopsin II, the protein sticks to another protein, called transducin. Before bumping into metarhodopsin II, transducin had tightly bound a small molecule called GDP. But when transducin interacts with metarhodopsin II, the GDP falls off, and a molecule called GTP binds to transducin. (GTP is closely related to, but critically different from, GDP.) (*Darwin's Black Box*, pp. 18-20)

Whew! There's a lot more in that vein from Behe, including descriptions of the cilia propulsion system in bacteria, the basic biochemistry of the immune system, and the cell's intricate internal transport system. Don't get the idea that *Darwin's Black Box* is a difficult read, though; the technical passages are set apart from the witty and graceful main text to facilitate skimming. Readers don't have to take in all the details to see the point, which is that Darwinian storytelling simply doesn't work at the molecular level. Each biochemical system requires a stupefyingly complex set of components which affect each other in intricate ways. No component makes sense except as part of the system, and the system doesn't work unless everything is in place. That's irreducible complexity.

It is notoriously difficult to prove a negative. No matter how irreducible the complexity seems, a storyteller can always invoke concepts like "preadaptation" to bolster the materialist faith that a Darwinian solution is somewhere out there. Fervent statements of faith aren't science, however, and fact-free science doesn't (usually) get published in biochemical journals. The key point in Behe's argument is that there are *no* papers in scientific journals that set out detailed, testable scenarios of how these incredibly complex biochemical systems could be produced by Darwinian-style processes. The very few papers that even attempt to speculate about this subject rely heavily on what scientists call "hand-waving." The journals of molecular evolution are full of papers documenting sequence comparisons, showing closer or more distant relationships between molecules. What they don't contain is papers documenting the existence of a Darwinian staircase up Mount Improbable. Until somebody fills the gap with scientific papers rather than stories, the best explanation for this situation is that the staircase doesn't exist.

Darwinism: Science or Philosophy?
Biochemists are not likely to challenge Behe in any fundamental way at the factual level. The scientific way to refute the irreducible complexity thesis is to publish papers detailing how the complex biochemi-

cal systems could have evolved, and the scientists already would have done that if they could. The controversy will not be over the facts but over whether Behe has gone "outside of science" by attributing irreducible complexity in biology to "design" rather than to some undiscovered material (i.e., mindless) mechanism.

Many scientists and philosophers think that a dedication to materialism is the defining characteristic of science. Their argument is that an a priori adherence to materialism is necessary to protect the very existence of science. If design in biology is real, then the Designer also might be real, and scientific materialists contemplate this possibility (if at all) with outright panic. Science will come to a screeching halt, they insist, because everybody will stop doing experiments and just attribute all phenomena to the inscrutable will of God.

Nonsense. On the contrary, the concept that the universe is the product of a rational mind provides a far better metaphysical basis for scientific rationality than the competing concept that everything in the universe (including our minds) is ultimately based in the mindless movements of matter. Perhaps materialism was a liberating philosophy when the need was to escape from dogmas of religion, but today materialism itself is the dogma from which the mind needs to escape. A rule that materialism should be professed *regardless of the evidence,* says Behe, is the equivalent of a rule that science may not contradict the teachings of a church. "It tries to place reality in a tidy box, but the universe will not be placed in a box."

Behe's fundamental principle is that "scientists should follow the physical evidence wherever it leads, with no artificial restrictions." Science has come as far as it has because scientists of the past were willing to describe the universe as it really is, rather than as the prejudices current in their times would have preferred it to be. The question is whether today's scientists have lost their nerve.

6

Daniel Dennett's
Dangerous Idea

DARWINISM IS NOT MERELY A BIOLOGICAL THEORY BUT A WAY OF
thinking about the world that generates powerful conclusions all the way
up and all the way down. In the academic world hardly anyone tries to
mount an overall challenge to Darwinism, but many distinguished think-
ers resist applying Darwinian logic to the subjects they know best. In
Darwin's Dangerous Idea *(Simon & Schuster, 1995) Daniel Dennett takes*
on the pussyfooters who flinch when Darwinism gets too close to home—
and in the following essay (first published in the October 1995 issue of
The New Criterion*) I take on Daniel Dennett.*

Daniel Dennett's fertile imagination is captivated by the very
dangerous idea that the neo-Darwinian theory of biological
evolution should become the basis for what amounts to an
established state religion of scientific materialism. Dennett takes the
scientific part of his thesis from the inner circle of contemporary
Darwinian theorists: William Hamilton, John Maynard Smith, George
C. Williams and the brilliant popularizer Richard Dawkins. When

Dennett describes the big idea emanating from this circle as dangerous, he does not mean that it is dangerous only to religious fundamentalists. The persons he accuses of flinching when faced with the full
implications of Darwinism are scientists and philosophers of the
highest standing: Noam Chomsky, Roger Penrose, Jerry Fodor, John
Searle and especially Stephen Jay Gould.

Each one of these very secular thinkers supposedly tries, as simple
religious folk do, to limit the all-embracing logic of Darwinism.
Dennett describes Darwinism as a "universal acid; it eats through just
about every traditional concept and leaves in its wake a revolutionized
world-view." One thinker after another has tried unsuccessfully to find
some way to contain this universal acid, to protect *something* from its
corrosive power. Why? First let's see what the idea is.

Dennett begins the account with John Locke's late-seventeenth-
century *Essay Concerning Human Understanding,* where Locke answers the question "Which came first, mind or matter?" Locke's
answer is that mind had to come first, because "it is impossible to
conceive that ever bare incogitative matter should produce a thinking
intelligent Being." David Hume mounted some powerful skeptical
arguments against this mind-first principle, but in the end he couldn't
come up with a solid alternative.

Darwin did not set out to overturn the mind-first picture of reality
but to do something much more modest: to explain the origin of
biological species and the wonderful adaptations that enable those
species to survive and reproduce in diverse ways. The answer
Darwin came up with was that these adaptations, which had seemed
to be intelligently designed, are actually products of a mindless
process called natural selection. Dennett says that what Darwin
offered the world, in philosophical terms, was "a scheme for creating Design out of Chaos without the aid of Mind." When the
Darwinian outlook became accepted throughout the scientific
world, the stage was set for a much broader philosophical revolution. Dennett explains:

Darwin's idea had been born as an answer to questions in biology, but it threatened to leak out, offering answers—welcome or not—to questions in cosmology (going in one direction) and psychology (going in the other direction). If [the cause of design in biology] could be a mindless, algorithmic process of evolution, why couldn't that whole process itself be the product of evolution, and so forth *all the way down?* And if mindless evolution could account for the breathtakingly clever artifacts of the biosphere, how could the products of our own "real" minds be exempt from an evolutionary explanation? Darwin's idea thus also threatened to spread *all the way up,* dissolving the illusion of our own authorship, our own divine spark of creativity and understanding. (p. 63)

The metaphysical reversal was so complete that it soon became as unthinkable within science to *credit* any biological feature to a Designer as it had previously been unthinkable to do without the Designer. Whenever seemingly insuperable problems were encountered—the genetic mechanism, the human mind, the ultimate origin of life—biologists were confident that a solution of the Darwinian kind would be found. To be sure, the cause of materialist reductionism was sometimes set back by "greedy reductionists" like the behaviorist B. F. Skinner, who tried to explain human behavior as a direct consequence of material forces. The catchy metaphor Dennett employs to describe the difference between greedy and good kinds of reductionism is "cranes, not skyhooks." The origin of (say) the human mind must be attributed to some process firmly anchored on the solid ground of materialism and natural selection (a crane) and not to a mystery or miracle (skyhook), but this does not mean human behavior or mental activity can be understood *directly* on the basis of material concepts like stimulus and response or natural selection.

Although many aspects of evolutionary theory remain controversial, Dennett asserts confidently that the overall success of Darwinism-in-principle has been so smashing that the basic program—all the way up and all the way down—is established beyond question. And yet the resistance continues. Some of it comes from religious people, who want to preserve some role for a Creator. Dennett just brushes aside the outright creationists but takes more pains to refute those who would say God is the author of the laws of nature, including that marvelous evolutionary process that does all the designing.

The Darwinian alternative to a Lawgiver at the beginning of the universe is to postpone the beginning indefinitely by hypothesizing something like an eternal system of evolution at the level of universes. For example, the physicist Lee Smolin has proposed that black holes are in effect the birthplaces of offspring universes, in which the fundamental physical constants would differ slightly from those in the parent universe. Since those universes that happened to have the most black holes would leave the most "offspring," the basic Darwinian concepts of mutation and differential reproduction could be extended to cosmology. Dennett contends that whether this or any other model is testable, at least cosmic Darwinism relies on the same kind of thinking that has been successful in scientific fields like biology where testing is possible, and that is enough to make it preferable to an alternative that brings in a skyhook. He does not attempt to explain the origin of the cosmic evolutionary process. It's just mutating universes all the way down.

Much of the resistance to Darwinism "all the way up" comes from scientists and philosophers who deny the capacity of natural selection to produce specifically human mental qualities like the capacity for language. Foremost among these is Noam Chomsky, founder of modern linguistics, who describes a complex language program seemingly "hard-wired" into the human brain, which has no real analogy in the animal world and for which there is no very plausible story of step-by-step evolution through adaptive intermediate forms. Chomsky readily accepts evolutionary naturalism in principle, but (supported by Stephen Jay Gould) he regards Darwinian selection as no more than a place holder for a true explanation, which has not yet been found, of the human language capacity.

To true-believing Darwinists like Richard Dawkins and Daniel

Dennett, all such objections are fundamentally misconceived. The more intricately "designed" a feature appears to be, the *more* certain it is to have been constructed by natural selection—because there is no alternative way of producing design without resorting to impossible skyhooks. Even in the toughest cases, where plausible Darwinian hypotheses are hard to imagine and impossible to confirm, a Darwinian solution simply has to be out there waiting to be found. The alternative to natural selection is either God or chance. The former is outside of science and also apparently outside the contemplation of Gould or Chomsky; the latter is no solution at all. Once you understand the dimensions of the problem, and the philosophical constraints within which it must be solved, Darwinism is practically true by definition—regardless of the evidence.

I call this a very interesting situation. Within science the Darwinian viewpoint clearly occupies the high ground, because nobody has come up with an alternative for explaining design that does not invoke an unacceptable preexisting Mind. (Dennett easily refutes such hype-induced notions as that a physics of self-organizing systems from the Santa Fe Institute is in the process of replacing Darwinism.) But the rulers of this impregnable citadel are worried because not everybody believes their citadel is impregnable. They are troubled not only by polls showing that the American public still overwhelmingly favors some version of supernatural creation, but also by the tendency of prominent scientists to accept Darwinism-in-principle but to dispute the theory's applicability to specific problems, usually the problems about which they are best qualified to speak.

Dennett thinks the dissenters either fail to understand the logic of Darwinism or shrink from embracing its full metaphysical implications. I prefer another explanation: Darwinism is a lot stronger as philosophy than it is as empirical science. If you aren't willing to challenge the underlying premise of scientific materialism, you are stuck with Darwinism-in-principle as a creation story until you find something better, and it doesn't seem that there *is* anything better.

Once you get past the uncontroversial examples of microevolution, however, such as finch-beak variations, peppered-moth coloring and selective breeding, all certainty dissolves in speculation and controversy. Nobody really knows how life originated, where the animal phyla came from or how natural selection could have produced the qualities of the human mind. Ingenious hypothetical scenarios for the evolution of complex adaptations are presented to the public virtually as fact, but skeptics within science derisively call them "just-so" stories, because they can be neither tested experimentally nor supported by fossil histories.

Many scientists who swear fealty to Darwinism on philosophical grounds put it aside when they get down to scientific practice. A good example is Niles Eldredge, a paleontologist who collaborated with Stephen Jay Gould in the famous papers proposing that evolution proceeds by "punctuated equilibria," meaning long, changeless periods that are occasionally interrupted by the abrupt appearance of new forms. "Punk eek" was widely interpreted at first as an implied endorsement of a macromutational alternative to Darwinian gradualism, a misunderstanding that led scornful Darwinists to dismiss the idea as "evolution by jerks," but both Gould and Eldredge insisted that the unseen process of change was Darwinian. Eldredge in particular is so determined to wash away the taint of heresy that he has taken to describing himself as a "knee-jerk neo-Darwinian," a label that seems both to protest too much and to imply a willingness to overlook disconfirming evidence.

On the other hand, Eldredge rejects what he calls "ultra-Darwinism," the position of Dawkins and Dennett, on grounds that obscurely imply rejection of the very factor that makes Darwin's idea dangerous, the claim that natural selection has sufficient creative power to account for design. For example, he writes in his 1994 book *Reinventing Darwin* that ultra-Darwinians are guilty of "physics envy" because they "seek to transform natural selection from a simple form of record keeping . . . to a more dynamic, active force that molds and shapes

organic form as time goes by." Eldredge has no philosophical problem
with atheistic materialism; his ambivalence stems entirely from the
embarrassingly un-Darwinian fossil record, as described in this typical paragraph:

> No wonder paleontologists shied away from evolution for so long. It
> never seems to happen. Assiduous collecting up cliff faces yields
> zigzags, minor oscillations, and the very occasional slight accumula-
> tion of change—over millions of years, at a rate too slow to account
> for all the prodigious change that has occurred in evolutionary history.
> When we do see the introduction of evolutionary novelty, it usually
> shows up with a bang, and often with no firm evidence that the fossils
> did not evolve elsewhere! Evolution cannot forever be going on some-
> where else. Yet that's how the fossil record has struck many a forlorn
> paleontologist looking to learn something about evolution. (*Reinvent-
> ing Darwin,* p. 95)

Whatever is motivating Eldredge to give all that fervent lip service to
Darwinism, it obviously is not anything he has discovered as a
paleontologist. In fact the real problem is understood by everyone,
although it has to be discussed in guarded terms. What paleontologists
fear is not the scientific consequences of disowning Darwinism but
the *political* consequences. They fear it might lead to a takeover of
government by religious fundamentalists, who might shut off the
funding for evolutionary science.

There are paleontologists who are more supportive of Darwinism
than Eldredge, just as there are other eminent scientists who are more
explicit in insisting that the neo-Darwinian variety of evolution is valid
only at the "micro" level. Regardless of the number or status of the
skeptics, the usual scientific practice is to retain a paradigm, however
shaky, until somebody provides a better one. I will assume *arguendo*
that this "best we've got" policy is justifiable within science itself. The
question I want to pursue is whether nonscientists have some legal,
moral or intellectual obligation to accept Darwinism as absolutely
true, especially when the theory is encountering so many difficulties

with the evidence. The issue comes up in many important contexts; here are two examples.

First, consider the situation of Christian parents, not necessarily fundamentalists, who suspect that the term *evolution* drips with atheistic implications. The whole point of Dennett's thesis is that the parents are dead right about the implications and that science educators who deny this are either misinformed or lying. Do parents then have a right to protect their children from indoctrination in atheism and even to insist that public schools include in the science curriculum a fair review of the arguments *against* the atheistic claim that unintelligent natural processes are our true creator?

Dennett cannot be accused of avoiding the religious liberty issue or of burying it in tactful circumlocutions. He proposes that theistic religion should continue to exist only in "cultural zoos," and he says this directly to religious parents:

> If you insist on teaching your children falsehoods—that the earth is flat, that "Man" is not a product of evolution by natural selection—then you must expect, at the very least, that those of us who have freedom of speech will feel free to describe your teachings as the spreading of falsehoods, and will attempt to demonstrate this to your children at our earliest opportunity. Our future well-being—the well-being of all of us on the planet—depends on the education of our descendants. (pp. 519-20)

Of course it is not freedom of speech that worries parents but the power of atheistic materialists to use public education for indoctrination, while excluding any other view as "religion." If you want to know how such threats sound to Christian parents, try imagining what would happen if some prominent Christian fundamentalist addressed similar language to Jewish parents. Would we think the Jewish parents unreasonable if they interpreted "at the very least" to imply that "at the very most" young children may be forcibly removed from the homes of recalcitrant parents and that those metaphorical cultural zoos may one day be enclosed by real barbed wire? Strong measures might seem justified if the well-being of everyone on the planet depends on protecting children from the falsehoods their parents want to tell them.

I will pass over the legal issues raised by this program of forced religious conversion, because the intellectual issues are even more interesting. Granted that Darwinism is the reigning paradigm in biology, is there some rule in the academic world that requires nonscientists to accept Darwinian principles when they write about, say, philosophy or ethics? My Berkeley colleague John Searle thinks so. In the first chapter of his recent book *The Construction of Social Reality,* Searle states that it is necessary "to make some substantive presuppositions about *how the world is in fact* in order that we can even pose the questions we are trying to answer (about how other aspects of reality are socially constructed)." According to Searle, "two features of our conception of reality are not up for grabs. They are not, so to speak, optional for us as citizens of the late twentieth and early twenty-first century." The two compulsory features are that the world consists entirely of the entities that physicists call particles and that living systems (including humans and their minds) evolved by natural selection.

I think Searle undermines his whole project by virtually ordering his readers not to notice that scientific materialism and Darwinism are themselves questionable philosophical doctrines rather than objective facts. Scientists assume materialism because they define their enterprise as a search for the best materialist theories, and this culturally driven methodological choice is not even evidence, let alone proof, that the world really does consist only of particles. As an explanation for design in biology, Darwinism is perfectly secure when it is regarded as a deduction from materialism but remarkably insecure when it is subjected to empirical testing. Given that what we most respect about science is its fidelity to the principle that empirical testing is what really matters, why should philosophers allow scientists to tell them that they must accept assumptions that don't pass the empirical test?

Searle is a particularly poignant example, because he is famous for defending the independence of the mind against the onslaught of the materialist "strong AI" program, and also for defending traditional academic standards against the corrosive relativism of the fact-value distinction. He is so skillful in argument that he almost holds his own even after leaping gratuitously into a pool of universal acid—but why accept the disadvantage? Searle could seize the high ground if he began by proposing that any true metaphysical theory must account for two essential truths that materialism cannot accommodate: first, that mind is more than matter; and second, that such things as truth, beauty and goodness really do exist even if most people do not know how to recognize them. Scientific materialists would answer that they proved long ago, or are going to prove at some time in the future, that materialism is true. They are bluffing.

Science is a wonderful thing in its place. Because science is so successful in its own territory, however, scientists and their allied philosophers sometimes get bemused by dreams of world conquest. Paul Feyerabend put it best: "Scientists are not content with running their own playpens in accordance with what they regard as the rules of the scientific method, they want to universalize those rules, they want them to become part of society at large, and they use every means at their disposal—argument, propaganda, pressure tactics, intimidation, lobbying—to achieve their aims." Samuel Johnson gave the best answer to this absurd imperialism. "A cow is a very good animal in the field; but we turn her out of a garden."

7

The Unraveling of Scientific Materialism

ALTHOUGH HARVARD GENETICS PROFESSOR RICHARD LEWONTIN IS NOT as well known to the public as Stephen Jay Gould or Richard Dawkins, he outranks both in the hierarchy of scientific professionals. He is highly reputed as an experimental scientist, sophisticated and well-informed in philosophical issues, and passionately leftist in politics. I respect him as an extraordinarily gifted man who, in embracing the faith of scientific materialism and Marxism, has bet his life on the wrong horse. I wonder if he is beginning to have his doubts. This essay first appeared in First Things, *November 1997.*

In a retrospective essay on Carl Sagan in *The New York Review of Books* (January 9, 1997), Richard Lewontin tells how he first met Sagan at a public debate in Arkansas in 1964. The two young scientists had been coaxed by senior colleagues to go to Little Rock to debate the affirmative side of this question: "RESOLVED, That the Theory of Evolution is as proved as is the fact that the Earth goes around the sun." Their main opponent was a biology professor from a fundamentalist college, with a Ph.D. from the University of Texas in zoology. Lewontin reports no details on the debate except to say that "despite our absolutely compelling arguments, the audience unaccountably voted for the opposition."

Of course Lewontin and Sagan attributed the vote to the audience's prejudice in favor of creationism. The resolution was framed in such a way, however, that the affirmative side should have lost even if the jury had been composed of Ivy League philosophy professors. How could the theory of evolution even conceivably be "proved" to the same degree as "the fact that the Earth goes around the sun"? The latter is an observable feature of present-day reality, whereas the former deals primarily with nonrepeatable events of the very distant past. The appropriate comparison would be between the theory of evolution and the accepted theory of the *origin* of the solar system.

If *evolution* referred only to currently observable phenomena like domestic animal breeding or finch-beak variation, then winning the debate should have been no problem for Lewontin and Sagan even with a fundamentalist jury. The statement "We breed a great variety of dogs," which rests on direct observation, is much easier to prove than the statement that the earth goes around the sun, which requires sophisticated reasoning. Not even the strictest biblical literalists deny dog breeding, finch-beak variations or similar instances of variation within a type. The more controversial claims of large-scale evolution are what arouse skepticism.

Scientists may think they have good reasons for believing that living organisms evolved naturally from nonliving chemicals or that complex organs evolved by the accumulation of micromutations through natural selection, but having reasons is not the same as having proof. I have seen people, previously inclined to believe whatever "science says," become skeptical when they realize that scientists actually seem to think that finch-beak or peppered-moth variation, or the mere existence of fossils, proves all the vast claims of "evolution."

It is as though the scientists, so confident in their answers, simply do not understand the question.

Carl Sagan described the theory of evolution in his final book as the doctrine that "human beings (and all the other species) have slowly evolved by natural processes from a succession of more ancient beings with no divine intervention needed along the way." It is the alleged absence of divine intervention throughout the history of life—that is, the strict *materialism* of the orthodox theory—that explains why a great many people, only some of whom are biblical fundamentalists, think Darwinian evolution (beyond the micro level) is basically materialistic philosophy disguised as scientific fact. Sagan himself worried about opinion polls showing that only about 10 percent of Americans believe in a strictly materialistic evolutionary process and, as Lewontin's anecdote concedes, some of the doubters have advanced degrees in relevant sciences. Dissent as widespread as that must rest on something less easily remedied than mere ignorance of facts.

Lewontin eventually parted company with Sagan over how to explain why the theory of evolution seems so obviously true to mainstream scientists and so doubtful to much of the public. Sagan attributed the persistence of unbelief to ignorance and hucksterism and set out to cure the problem with popular books, magazine articles and television programs promoting the virtues of mainstream science over its fringe rivals. Lewontin, a Marxist whose philosophical sophistication exceeds that of Sagan by several orders of magnitude, came to see the issue as essentially one of basic philosophical commitment rather than factual knowledge.

The reason for opposition to scientific accounts of our origins, according to Lewontin, is not that people are ignorant of facts but that they have not learned to think from the right starting point. In his words, "The primary problem is not to provide the public with the knowledge of how far it is to the nearest star and what genes are made of. . . . Rather, the problem is to get them to reject irrational and supernatural explanations of the world, the demons that exist only in their imaginations, and to accept a social and intellectual apparatus, Science, as the only begetter of truth." What the public needs to learn is that, like it or not, "we exist as material beings in a material world, all of whose phenomena are the consequences of material relations among material entities." In a word, the public needs to accept materialism, which means that they must put God (whom Lewontin calls the "Supreme Extraterrestrial") in the trash can of history where he belongs.

Although Lewontin wants the public to accept science as the only source of truth, he freely admits that mainstream science itself is not free of the hokum that Sagan so often found in fringe science. As examples he cites three influential scientists who are particularly successful at writing for the public—E. O. Wilson, Richard Dawkins and Lewis Thomas—

> each of whom has put unsubstantiated assertions or counter-factual claims at the very center of the stories they have retailed in the market. Wilson's *Sociobiology* and *On Human Nature* rest on the surface of a quaking marsh of unsupported claims about the genetic determination of everything from altruism to xenophobia. Dawkins's vulgarizations of Darwinism speak of nothing in evolution but an inexorable ascendancy of genes that are selectively superior, while the entire body of technical advance in experimental and theoretical evolutionary genetics of the last fifty years has moved in the direction of emphasizing non-selective forces in evolution. Thomas, in various essays, propagandized for the success of modern scientific medicine in eliminating death from disease, while the unchallenged statistical compilations on mortality show that in Europe and North America infectious diseases ... had ceased to be major causes of mortality by the early decades of the twentieth century.

Lewontin laments that even scientists frequently cannot judge the reliability of scientific claims outside their fields of specialty and have to take the word of recognized authorities on faith. "Who am I to believe about quantum physics if not Steven Weinberg, or about the solar system if not Carl Sagan? What worries me is that they may believe what Dawkins and Wilson tell them about evolution."

One major living scientific popularizer whom Lewontin does *not* trash is his Harvard colleague and political ally Stephen Jay Gould. Just to fill out the picture, however, it seems that admirers of Dawkins have as low an opinion of Gould as Lewontin has of Dawkins and Wilson. According to a 1994 *New York Review of Books* essay by John Maynard Smith, dean of British neo-Darwinists,

> the evolutionary biologists with whom I have discussed [Gould's] work tend to see him as a man whose ideas are so confused as to be hardly worth bothering with, but as one who should not be publicly criticized because he is at least on our side against the creationists. All this would not matter, were it not that he is giving non-biologists a largely false picture of the state of evolutionary theory.

Lewontin fears that nonbiologists will fail to recognize that Dawkins is peddling pseudoscience; Maynard Smith fears exactly the same with respect to Gould.

If eminent experts say that evolution according to Gould is too confused to be worth bothering about, and others equally eminent say that evolution according to Dawkins rests on unsubstantiated assertions and counterfactual claims, the public can hardly be blamed for suspecting that grand-scale evolution may rest on something less impressive than rock-solid, unimpeachable fact. Lewontin confirms this suspicion by explaining why "we" (that is, the kind of people who read *The New York Review of Books*) reject out of hand the view of those who think they see the hand of the Creator in the material world:

> We take the side of science *in spite of* the patent absurdity of some of its constructs, *in spite of* its failure to fulfill many of its extravagant promises of health and life, *in spite of* the tolerance of the scientific community for unsubstantiated just-so stories, because we have a prior commitment, a commitment to materialism. It is not that the methods and institutions of science somehow compel us to accept a material explanation of the phenomenal world, but, on the contrary, that we are forced by our *a priori* adherence to material causes to create an apparatus of investigation and a set of concepts that produce material

explanations, no matter how counterintuitive, no matter how mystify-
ing to the uninitiated. Moreover, that materialism is absolute, for we
cannot allow a Divine Foot in the door.

That paragraph is the most insightful statement of what is at issue in
the creation-evolution controversy that I have ever read from a senior
figure in the scientific establishment. It explains neatly how the theory
of evolution can seem so certain to scientific insiders and so shaky to
the outsiders. For scientific materialists *the materialism comes first;
the science comes thereafter.* We might therefore more accurately term
them "materialists employing science." And if materialism is true,
then some materialistic theory of evolution has to be true simply as a
matter of logical deduction, regardless of the evidence. That theory
will necessarily be at least roughly like neo-Darwinism, in that it will
have to involve some combination of random changes and lawlike
processes capable of producing complicated organisms that (in
Dawkins's words) "give the appearance of having been designed for
a purpose."

The a priori commitment to materialism explains why evolutionary
scientists are not disturbed when they learn that the fossil record does
not provide examples of gradual macroevolutionary transformation,
despite decades of determined effort by paleontologists to confirm
neo-Darwinian presuppositions. That is also why origin-of-life chem-
ists like Stanley Miller continue in confidence even when geochemists
tell them that the early earth did not have a reducing (oxygen-free)
atmosphere, essential for producing the chemicals required by the
prebiotic soup scenario. They reason that there had to be some source
(comets?) capable of providing the needed molecules, because otherwise life would not have evolved. When evidence showed that the
period available on the early earth for the evolution of life was
extremely brief in comparison to the time previously posited for
chemical evolution scenarios, Sagan calmly concluded that the chemical evolution of life must be easier than we had supposed, because it
happened so rapidly on the early earth.

That is also why neo-Darwinists like Richard Dawkins are not troubled by the Cambrian Explosion, where all the invertebrate animal groups appear suddenly and without identifiable ancestors. Whatever the fossil record may suggest, those Cambrian animals had to evolve by accepted neo-Darwinian means, which is to say by material processes requiring no intelligent guidance or supernatural input. Materialist philosophy demands no less. That is also why Niles Eldredge, surveying the absence of evidence for macroevolutionary transformations in the rich marine invertebrate fossil record, can observe that "evolution always seems to happen somewhere else" and then describe himself on the very next page as a "knee-jerk neo-Darwinist." Finally, that is why Darwinists do not take critics of materialist evolution seriously but speculate instead about "hidden agendas" and resort immediately to ridicule. In their minds, to question materialism is to question reality. All these specific points are illustrations of what it means to say that "we" have an a priori commitment to materialism.

Those in scientific leadership cannot afford to disclose that commitment frankly to the public. Imagine what chance the affirmative side would have if the question for public debate were rephrased candidly as "RESOLVED, that everyone should adopt an a priori commitment to materialism." Everyone would see what many now sense dimly: that a methodological premise which is useful for limited purposes has been expanded to form a metaphysical absolute.

Of course people who define science as the search for materialistic explanations will find it useful to assume that such explanations always exist. To suppose that a philosophical preference can validate a cherished theory is to define "science" as a way of supporting prejudice. Yet that is exactly what the Darwinists seem to be doing when their evidence is evaluated by critics who are willing to question materialism.

One of those critics, bearing impeccable scientific credentials, is Michael Behe. Behe argues that complex molecular systems (such as bacterial and protozoan flagella, immune systems, blood clotting and cellular transport) are "irreducibly complex." This means that the

systems incorporate elements that interact with each other in such
complex ways that it is impossible to describe detailed, testable
Darwinian mechanisms for their evolution. Never mind for now
whether you think that Behe's argument can prevail over sustained
opposition from the materialists. The primary dispute is not over who
is going to win but about whether the argument can even get started.
If we know a priori that materialism is true, then contrary evidence
properly belongs under the rug, where it has always been swept.

For Lewontin, the public's determined resistance to scientific materialism constitutes "a deep problem in democratic self-governance."
Quoting Jesus' words from the Gospel of John, he thinks "the truth that
makes us free" is not an accumulation of knowledge but a metaphysical
understanding (i.e., materialism) that sets us free from belief in supernatural entities like God. How is the scientific elite to persuade or
bamboozle the public to accept the crucial starting point? Lewontin turns
for guidance to the most prestigious of all opponents of democracy, Plato.
In his dialogue *Gorgias,* Plato reports a debate between the rationalist
Socrates and three Sophists, or teachers of rhetoric. The debaters all
agree that the public is incompetent to make reasoned decisions on
justice and public policy. The question in dispute is whether the
effective decision should be made by experts (Socrates) or by the
manipulators of words (the Sophists).

In familiar contemporary terms, the question might be stated as
whether a court should appoint a panel of impartial authorities to
decide whether the defendant's product caused the plaintiff's cancer
or whether the jury should be swayed by rival trial lawyers each
touting their own experts. Much turns on whether the authorities are
truly impartial or whether they have interests of their own. When the
National Academy of Sciences appoints a committee to advise the
public on evolution, it consists of persons picked in part for their
scientific outlook, which is to say their a priori acceptance of materialism. Members of such a panel (1) know a lot of facts in their specific
areas of research and (2) have a lot to lose if the "fact of evolution" is

exposed as a philosophical assumption. Should skeptics accept such persons as impartial fact-finders? Lewontin himself knows too much about cognitive elites to say anything so naive, and so in the end he gives up and concludes that "we" do not know how to get the public to the right starting point.

Lewontin is brilliantly insightful but too crankily honest to be as good a manipulator as his Harvard colleague Stephen Jay Gould. Gould displays both his talent and his unscrupulousness in his essay "Nonoverlapping Magisteria," whose title is followed by a summary blurb: "Science and religion are not in conflict, for their teachings occupy distinctly different domains" (*Natural History,* March 1997). With a summary statement like that, you can be sure that Gould is out to reassure the public that evolution leads to no alarming conclusions. True to form, Gould insists that the only dissenters from evolution are "Protestant fundamentalists who believe that every word of the Bible must be literally true." Gould also insists that evolution (he never defines the word) is "both true and entirely compatible with Christian belief." Gould is familiar with nonliteralist opposition to evolutionary naturalism, but he blandly denies that any such phenomenon exists. He even quotes a letter written to *The New York Times* in answer to an op-ed essay by Michael Behe, without revealing the context. You can do things like that when you know that the media won't call you to account.

The centerpiece of Gould's essay is an analysis of the complete text of Pope John Paul's statement of October 22, 1996, to the Pontifical Academy of Sciences, endorsing evolution as "more than a hypothesis." He fails to quote the pope's crucial qualification that "theories of evolution which, in accordance with the philosophies inspiring them, consider the spirit as emerging from the forces of living matter or as a mere epiphenomenon of this matter, are incompatible with the truth about man." Of course a theory based on materialism assumes by definition that there is no "spirit" active in this world that is independent of matter. Gould knows this perfectly well, and he also knows, just as Lewontin does, that the evidence doesn't support claims for the creative power of natural selection made by such as Richard Dawkins. That is why the philosophy that really supports the theory has to be protected from critical scrutiny.

Gould's essay is a tissue of half-truths aimed at putting religious people to sleep or luring them into a "dialogue" on terms set by materialists. Thus Gould graciously allows religion to participate in discussions of morality or the meaning of life, because science does not claim authority over such questions of value, and because "religion is too important to too many people for any dismissal or denigration of the comfort still sought by many folks from theology." (Translation: "Some people are too weak to get along without that crutch, and they vote. Better throw them a bone.") Gould insists, however, that all such discussion must cede to science the power to determine the *facts,* and one of the facts is an evolutionary process that is every bit as materialistic and purposeless for Gould as it is for Lewontin or Dawkins. If religion wants to accept a dialogue on those terms, it's fine with Gould—but don't let those religious people think they get to make an independent judgment about the evidence that supposedly supports the "facts." And if the religious people are gullible enough to accept materialism as one of the facts, they won't be capable of causing much trouble.

The creation-evolution debate is not deadlocked. Propagandists like Gould try to give the impression that nothing has changed, but essays like Lewontin's and books like Behe's demonstrate that honest thinkers on both sides are near agreement on a redefinition of the conflict. Biblical literalism is not the issue. The issue is whether materialism and rationality are the same thing. Darwinism is based on an a priori commitment to materialism, not on a philosophically neutral assessment of the evidence. Separate the philosophy from the science, and the proud tower collapses. When the public understands this clearly, Lewontin's Darwinism will start to move out of the science curriculum and into the department of intellectual history, where it can gather dust on the shelf next to Lewontin's Marxism.

8

The Gorbachev
of Darwinism

IN CHAPTER FOUR OF REASON IN THE BALANCE *I DESCRIBED THE VERY different theories of evolution held by the two leading popularizers of the subject, Richard Dawkins and Stephen Jay Gould. As Adrian Desmond wrote of the earlier differences between Darwin and Huxley, sometimes it seems that the only things they have in common are their belief in materialism and their hatred of creationism. The differences have been suppressed until now in the interest of keeping the Divine Foot out of the door, but at last the quarrel is coming out in public. Allies of Dawkins like John Maynard Smith seem to have lost patience after Gould published a wildly hostile review of a book by Dawkins protégé Helena Cronin,* The Ant and the Peacock, *in* The New York Review of Books *issue of November 19, 1992. Maynard Smith and Daniel Dennett subsequently took off the gloves, and eventually Gould had to reply in kind. This essay first appeared in* First Things, *January 1998.*

S tephen Jay Gould is mad as hell, and he's not going to take it anymore. Readers of *The New York Review of Books** learned that much in 1997, if they read a lengthy, two-part tirade in which Gould attempted to settle scores with some of his more prominent enemies within the guild of Darwinists. The specific targets were Daniel Dennett, John Maynard Smith, Robert Wright and especially, although largely in the background, Richard Dawkins. One cannot understand the controversy without sampling the level of vitriol, which may be judged by this salvo from Gould:

> [Dennett's] limited and superficial book reads like a caricature of a caricature—for if Richard Dawkins has trivialized Darwin's richness by adhering to the strictest form of adaptationist argument in a maximally reductionist mode, then Dennett, as Dawkins's publicist, manages to convert an already vitiated and improbable account into an even more simplistic and uncompromising doctrine. If history, as often noted, replays grandeurs as farces, and if T. H. Huxley truly acted as "Darwin's bulldog," then it is hard to resist thinking of Dennett, in this book, as "Dawkins's lapdog."

After going on in that vein for some pages, Gould responded with hurt feelings to Maynard Smith's published comment that "the evolutionary biologists with whom I have discussed [Gould's] work tend to see him as a man whose ideas are so confused as to be hardly worth bothering with, but as one who should not be publicly criticized because he is at least on our side against the creationists." To this Gould lamented that Maynard Smith used to say much nicer things about him, and he piously warned that "we will not win this most important of all battles [against the creationists] if we descend to the

---

*Stephen Jay Gould, "Darwinian Fundamentalism," *The New York Review of Books,* June 12, 1997, p. 34; and "Evolution: The Pleasures of Pluralism," *The New York Review of Books,* June 26, 1997, p. 47. Letters from Daniel Dennett and Robert Wright appear in the August 14 issue, p. 64, with Gould's scorching reply to both. Finally a bitter exchange with Steven Pinker appeared in the issue of October 9, 1997. Overall, it's the intellectual equivalent of a barroom brawl. All these articles and exchanges may be read at the *New York Review*'s web site, http://www.nybooks.com/nyrev/

same tactics of backbiting and anathematization that characterize our true opponents." Tell that to Dawkins's lapdog.

Gould's decision to publish an all-out blast at the writers whom he calls "Darwinian Fundamentalists" escalated what his colleague Niles Eldredge has called the "high table debate" among evolutionists. This is basically a struggle between classical neo-Darwinists (represented most prominently by Dawkins) and revisionists (headed by Gould himself) who follow the tradition of T. H. Huxley by advocating "evolution" while remaining cool toward Darwin's distinctive mechanism. It's a debate that has long been muted because of the adversaries' mutual desire to avoid giving ammunition to the despised creationists, and even now the arguments are conducted in an obscure jargon worthy of Pravda in its heyday. So here's what it's all about.

In the early 1980s the British geneticist J. R. G. Turner remarked, with specific reference to the controversies swirling around Gould, that "evolutionary biologists are all Darwinists, as all Christians follow Christ and all Communists, Karl Marx. The schisms are over which parts of the Master's teaching shall be seen as central." The canonical text for fans of natural selection is Darwin's eloquent statement in *On the Origin of Species* that

> natural selection is daily and hourly scrutinising, throughout the world, every variation, even the slightest; rejecting that which is bad, preserving and adding up all that is good; silently and insensibly working, whenever and wherever opportunity offers, at the improvement of each organic being in relation to its organic and inorganic conditions of life.

Note the key elements: natural selection everywhere and at all times accepts or rejects *all* variations, however slight, and continually promotes the "improvement" of all organisms. Evolution of that kind, in the jargon of the trade, is called panselectionism.*

---

*The quotations in this paragraph are from John R. G. Turner, " 'The Hypothesis That Explains Mimetic Resemblance Explains Evolution': The Gradualist-Saltationist Schism," in *Dimensions of Darwinism*, ed. Marjorie Greene (Cambridge University Press, 1983), pp. 129-69, and

The revisionist Gould calls that picture of ubiquitous selection "ultra-Darwinism" or "Darwinian fundamentalism," and he attributes it not to Darwin himself but to contemporary Darwinists like Dawkins and Dennett. Gould ignores Darwin's own panselectionist affirmations and quotes instead a passage from the sixth and final (1872) edition of the *Origin*. There Darwin remarked with some bitterness that critics had, by "steady misrepresentation," overlooked his qualification that "natural selection has been the main *but not the exclusive* means of modification" (emphasis added). Whether the qualification amounts to much is hard to say, since a few exceptions to an otherwise pervasive pattern of selectionism would be consistent with the modest disclaimer that natural selection is not literally "exclusive."

In any event, Gould accuses the ultra-Darwinists of preaching that "natural selection regulates everything of any importance in evolution, and that adaptation emerges as a universal result and ultimate test of selection's ubiquity." Against this fundamentalism Gould asserts his own "pluralism," which includes at least four nonadaptationist aspects of evolution: (1) neutral genetic changes are a major aspect of evolution; (2) basic developmental pathways are highly conserved across otherwise disparate groups and hence impose constraints on adaptive change; (3) species remain unchanged for long periods and then branch apart in "geological moments" (the jargon for this is "punctuated equilibria"); and (4) many or most extinctions have been due to catastrophic events rather than (as Darwin insisted) to the gradual operation of ordinary selective pressures.

That's where the name-calling starts, because classical Darwinists consider Gould's description of their position a preposterous caricature. Gould has a well-earned reputation for distorting the views of his rivals and adversaries, and so it is not surprising to find

---

from Charles Darwin, *On the Origin of Species* (Penguin ed., 1982), p. 133.

that the complaints are justified. To my knowledge none of his targets disputes that neutral variations occur in plenty, that developmental pathways are conserved, that significant evolutionary change may occur in brief periods of time (geologically speaking) after longer periods of stasis, or that the dinosaurs were probably wiped out by a planetary catastrophe. Gould does deserve credit for advocating these subtheories before they became popular, but nowadays everybody claims to be a pluralist.

For his own part, Gould does not deny the central tenet of the classicists—that adaptive complexity is due to the Darwinian mechanism of natural selection. In his own words, "Yes, eyes are for seeing and feet are for moving. And, yes again, I know of no scientific mechanism other than natural selection with the proven power to build structures of such eminently workable design." The creative power of natural selection is actually inferred from materialist philosophy rather than proved by scientific evidence, but let that pass. If both sides agree that natural selection is responsible for adaptation and also that natural selection isn't the whole story of evolution, then what's the beef? Little wonder that many observers have concluded that there is no substance behind this food fight at the high table, only a clash of overgrown egos.

In fact, however, the controversy is substantive. The key to understanding the substance is to recognize that being a true Darwinist requires more than giving lip service to natural selection before going on to something else, which is what Gould typically does. If natural selection actually made all those marvels of biological complexity, certain conclusions about the pace and manner of evolution necessarily follow, and Gould frequently seems to be denying those necessary conclusions. The dinosaurs can be killed off as rapidly as you like, but all the dinosaurs that died and all the new mammals that replaced them had to have been built up in the first place through the gradual accumulation of random mutations by natural selection. Likewise, the problem with neutral gene substitutions is not that anyone doubts that they occur but that neutral changes by definition don't help with the overwhelming task of building up the complex adaptations. Natural selection had to do that whole job, if God didn't do it, and that means natural selection had to be continuously active across vast stretches of geological time—regardless of what the fossil record may or may not show. That implies, among other things, that an enormous amount of evidence of the process has to be missing from the fossil record, and yet Gould frequently gives the impression that he thinks the evidence was never there.

One of the most notorious examples occurs in Gould's discussion of the Cambrian Explosion in his book *Wonderful Life*. The Cambrian Explosion is the sudden appearance of the major animal groups (phyla) in the rocks of the Cambrian era, without apparent ancestors. As Dawkins himself has put it, "It is as though they were just planted there, without any evolutionary history." Of course Dawkins and all other Darwinists believe that this appearance is an illusion caused by the incompleteness of the record and that a complete fossil record would show a universe of transitional forms and side branches, all having evolved by tiny steps from a single common ancestor. Gould raises a radically different possibility. He explains that there are two possible explanations for the absence of Precambrian ancestors: "the artifact theory (they did exist, but the fossil record hasn't preserved them), and the fast-transition theory (they really didn't exist, at least as complex invertebrates easily linked to their descendants).*

That final qualifying clause is a typical example of Gould's penchant for equivocation: *of course* the missing ancestors didn't exist in

---

*Richard Dawkins, *The Blind Watchmaker* (Longman, 1986), p. 229; Stephen Jay Gould, *Wonderful Life* (Norton, 1989), pp. 271-72. Dawkins chided Gould in the professional literature by writing, "Even if there really was a Cambrian explosion such that all the major phyla diverged during a 10-million-year period, this is no reason to think that Cambrian evolution was a qualitatively special kind of super-jumpy process. Bauplane [body plans] don't drop out of a clear Platonic sky: they evolve step by step from predecessors, and they do so (I bet, and so would Gould if explicitly challenged) under approximately the same Darwinian rules as we see today." Richard Dawkins, "Human Chauvinism," *Evolution* 51 (1997): 1019. Dawkins is obviously right—*if* Darwinism is true.

a form "easily linked to their descendants." That is why there is a problem, and why the artifact theory has to be true if Darwinism is true. Hence when Gould went on to proclaim that new discoveries had sounded "the death knell of the artifact theory," some readers understandably took him to be saying that the phyla really *were* just planted there without any evolutionary history, which amounts to saying that they were specially created!

Gould assuredly couldn't have meant *that,* but then what exactly *did* he mean? Remember that saving Darwinism in the teeth of the Cambrian evidence requires not just assuming a few missing ancestors, easily linked to their descendants or not, but assuming a vast quantity of vanished transitional forms between the hypothetical single-celled ancestors and the vastly different multicellular invertebrates. If you are a Darwinist you *know* the necessary ancestors and transitionals had to exist, regardless of the lack of fossil evidence. If you doubt that their absence is an artifact of the fossil record, you are not a Darwinist.

The difficulty of saying whether Gould really is a Darwinist or not stems from his habit of combining radically anti-Darwinian statements with qualifications that preserve a line of retreat. When Gould loudly proclaimed "the return of the hopeful monster," for example, he seemed to be endorsing geneticist Richard Goldschmidt's view that large mutations create new kinds of organisms in single-generation jumps—a heresy that Darwinists consider to be only a little better than outright creationism. If you read the fine print carefully, however, you'll find that Gould surrounded his claims with qualifications allowing him to insist that he is at least somewhere in the outfield of the ballpark of orthodoxy. Even when Gould bluntly announced that neo-Darwinism is "effectively dead," it turns out that he only meant— well, nobody seems to know what he meant, but certainly not that neo-Darwinism is effectively dead.

For years Darwinists like Maynard Smith gave Gould the benefit of the doubt, appreciating his genuine flair and his willingness to fight the common enemy. At last they have become thoroughly exasperated with his "now you see it, now you don't" practice of vaguely affirming Darwinism while specifically denying its necessary implications. Gould will only have exacerbated their disgust with his latest outburst.

Gould's uncomfortable situation reminds me of the self-created predicament of Mikhail Gorbachev in the last years of the Soviet empire. Gorbachev recognized that something had gone wrong with the communist system, but he thought the system itself could be preserved if it was reformed. His democratic friends warned him that Marxist fundamentalists would inevitably turn against him, but he was unwilling to endanger his position in the ruling elite by following his own logic to its necessary conclusion. Gould, like Gorbachev, deserves immense credit for bringing glasnost to a closed society of dogmatists. And like Gorbachev, he lives on as a sad reminder of what happens to a potentially great man who lacked the nerve to make a clean break with a dying theory.

9

A Metaphysics Lesson

THE PUBLIC DEBATE OVER CREATION AND EVOLUTION TOOK A radically new turn in 1997, as newspaper and television reporters began to recover from their addiction to the "Inherit the Wind stereotype." (See chapter two of Defeating Darwinism by Opening Minds *[InterVarsity Press, 1997].) Dealing with media scrutiny required Darwinists to put a theistic spin on their materialism, and they rose to the occasion. This essay is previously unpublished.*

In 1995 the U.S. National Association of Biology Teachers (NABT) promulgated a "Statement on Teaching Evolution" to guide high-school teachers in handling this delicate subject. The most significant part of the statement proclaimed:

> The diversity of life on earth is the outcome of evolution: an unsupervised, impersonal, unpredictable and natural process of temporal descent with genetic modification that is affected by natural selection, chance, historical contingencies and changing environments.

Because of its official status, the statement was used by critics of evolutionary naturalism, including me, to show what biology textbooks mean when they tell students that "evolution is a fact." The textbooks emphatically do *not* mean that evolution is a God-guided process or that our existence might be the product of a divine plan rather than an accident of a material universe. Note that the quoted sentence makes this point both negatively and positively. Evolution is *not* supervised, or personal, or predictable, or reflective of anything supernatural. It *is* affected by various natural factors but (by negative implication) not by any intelligent causes or purposeful forces. All this is standard Darwinian metaphysical naturalism, but in official documents it is usually pervasively insinuated rather than baldly stated in such an "in your face" manner.

The statement did make a halfhearted effort to deny that the NABT was saying anything about religion: "Evolutionary theory, indeed all of science, is necessarily silent on religion and neither refutes nor supports the existence of a deity or deities." The total effect was something like that produced by one of those nondenials that politicians issue just before resigning from office. "We are not saying that God does not *exist;* we are just saying that God had nothing to do with the natural, unsupervised process that created us."

Two distinguished scholars of religion and philosophy, Professors Huston Smith and Alvin Plantinga, wrote a letter to the NABT board of directors suggesting that the words *unsupervised* and *impersonal* were inappropriate because "science presumably doesn't address such theological questions, and isn't equipped to deal with them. How could an empirical inquiry possibly show that God was not guiding and directing evolution?" The Smith/Plantinga letter went on to say that the statement "gives aid and comfort to extremists in the religious right for whom it provides a legitimate target" and that "because of its logical vulnerability, it lowers Americans' respect for scientists and their place in our culture." Therefore deletion of the two words "would help defuse tensions which are causing unnecessary problems in our collective life."

The NABT board considered the Smith/Plantinga plea in October 1997 and initially voted unanimously to reject it. Wayne Carley, NABT's executive director, explained to reporters that the directors felt "rather strongly" that they should make no change in the statement, because "we believe it," and because "altering the statement would give creationists "just the aid and comfort Smith and Plantinga argue against." A few days later, however, the board reversed itself and voted to remove the two offending words—also unanimously. The prime mover in the reversal was Eugenie Scott, executive director of the National Center for Science Education (NCSE), a private organization dedicated to protecting the teaching of evolution from creationist challenges. Scott warned the board that it was in danger of giving the public the impression that "evolution" effectively means atheism. As she later explained on the NCSE web page,

> After the Statement was published, antievolutionists criticized the use of the terms "unsupervised" and "impersonal." UC Berkeley lawyer Phillip Johnson (author of *Darwin on Trial*) and other antievolutionists have claimed that the NABT statement is "proof" that evolution is inherently an antireligious ideological system, rather than simply a well-supported scientific explanation. Criticisms of the NABT statement have appeared in newspaper letters to the editor, newsletters and other publications. It appears that when most Americans other than scientists hear evolution described in blanket fashion as "unsupervised" they hear "God had nothing to do with it"—a statement which is outside of what science can tell us.

According to the *Chicago Tribune,* Scott (herself an agnostic) said she asked board members, " 'Would the statement mean the same thing if it said evolution means you can't believe in God and life is meaningless?' They all kind of turned pale." Explaining why the board changed its mind, Wayne Carley told the *Cleveland Plain Dealer* that the statement "was interpreted to mean we were saying there is not God. Absolutely not. We did not mean to imply that. That's beyond the purview of science. . . . We had only intended to say there is no evidence the process of evolution is directed from some source."* Scott reassured her web-page readers that

> evolution is still described as a "natural process" (the only phenomena science can study), and a later [sentence] states that natural selection "has no specific direction or goal, including survival of a species." The strong position of evolution in biology and other sciences was uncompromised by removing two adjectives that miscommunicated NABT's meaning.

In short, nothing was changed. Two particularly attention-getting words had been dropped, but the rest of the statement still carried the same message.

The New York Times had a different impression, however. The opening lines of a story, prominently displayed on the front page of its Sunday "Week in Review" section, portrayed the NABT reversal as a "startling about-face" which was "clearly designed to allow for the possibility that a Master Hand was at the helm." The bulk of the story concerned a group of "new creationists"—including Phillip Johnson, Michael Behe and David Berlinski—who were portrayed as making headway in moving the debate over creation and evolution into the intellectual mainstream. The *Times* cited the NABT reversal as tangible proof of their success: "This surprising change in creed for the nation's biology teachers is only one of many signs that the proponents of creationism, long stereotyped as anti-intellectual Bible-thumpers, have new allies and the hope of new credibility."**

---

*See Steve Kloehn, "On Second Thought, Biology Teachers Leave Room for God," *Chicago Tribune,* October 17, 1997, p. 12; Ira Rifkin, "Teachers Change Evolution Wording," *Cleveland Plain Dealer,* October 16, 1997, p. 10E.

**See Laurie Goodstein, "Christians and Scientists: New Light for Creationism," *The New York Times,* December 21, 1997, Week in Review section, p. 1. The occasion for the *Times* story was a special two-hour debate on William F. Buckley's national PBS TV program, *Firing Line,* on creation and evolution. In the debate the evolution team, which included Eugenie Scott, labored mightily to give the impression that the "evolution" science teachers are promoting is a God-guided process. Their "team captain," attorney Barry Lynn, went so far as to quote John 1:1 ("In the beginning was the Word"), to suggest that the "Word" was *Evolve.* This strategy

The New York Times story was reprinted in dozens of newspapers nationwide, to the horror of NABT officials and sympathizers, who protested to the *Times* that the story had given "the impression that the NABT had capitulated to creationists." The *Times* editor responded sternly to these complaints by e-mail: "The revision of your platform ... clearly allows for the possibility that evolution was guided by some omniscient power. And that, not the *Times* article, is what may give creationists comfort, wrongheaded though that may seem to you." The NABT president gamely insisted to the *Denver Post* that "there was no backing down" but admitted, "We knew hard-core creationists would regard this as a victory, and they have."

The NABT got into all this trouble because its original statement came too close to saying explicitly, "God is not your Creator." Such a statement either injects the NABT into the domain of religion or brings God into the domain of science. The Smith/Plantinga letter asked the pertinent question: "How could an empirical inquiry possibly show that God was not guiding and directing evolution?"

If empirical inquiry cannot make that negative showing, then the NABT statement went beyond scientific evidence to advocate what amounts to atheism. If that is what biology teachers are doing when they promote "evolution," then they are inviting political and legal opposition. American constitutional law requires that public schools be "neutral" on religious questions, and atheism is hardly a neutral position. Religious conservatives are convinced that the public school curriculum pervasively insinuates that God is either nonexistent or unworthy of attention, and the Smith/Plantinga letter warned the NABT that its statement virtually conceded that they are right.

An even more serious problem would be created if the NABT meant

---

worked well for the Darwinists during the debate, but I was sure it would cause trouble for them down the road. Double-talk and spin-doctoring will sometimes gain an advantage in the short run, but in the long run a science establishment that relies on obfuscation is in serious trouble. For the *Denver Post*'s follow-up story, see Cate Terwilliger, "Changes in Biology Teachers' Platform Rekindles Creationism Fire," *Denver Post*, January 29, 1998, p. E1.

that empirical inquiry *can* show that God did not guide and direct evolution. That claim makes God's guidance (or intelligent causation) a falsifiable hypothesis within science. Whatever empirical evidence can negate it can also support. For example, Michael Behe (see essay five in this book, "The Storyteller and the Scientist") has argued that various functional molecular systems in biology are irreducibly complex and hence cannot be assembled without the participation of an intelligent agent. The NABT is among the scientific organizations that have fought to keep that possibility off the table by labeling it "religion, not science." But if the presence or absence of intelligent causes in biology is testable, then intelligent design is a legitimate scientific hypothesis. Of course materialists would claim that intelligent design is a false hypothesis and that natural selection and similar unintelligent causes are perfectly capable of doing all the creating, but they know very well that many people would evaluate the evidence differently.

The NABT directors did not at first perceive the dilemma, because the problem does not exist for people who think within the metaphysical categories of scientific materialism. What the NABT meant to say all along is that there can be no *scientific evidence* that an intelligent agent played a role in evolution. An invisible agent might have exerted an undetectable influence—science can't negate a possibility like that. What is out of bounds is any claim that there is evidence for such an agent—*because science in our culture excludes that possibility by definition.* Behe may be right that molecular systems seem irreducibly complex, but that merely means that science has not completed its work of explanation. Natural processes that produced the complexity will inevitably be discovered. Similarly, critics of prebiological evolution may be right that science is a long way from discovering how life originated, but that is a temporary discouragement that will be overcome any day by a breakthrough. Opponents of reductionism may be right for now in insisting that science has no explanation for human consciousness and free will, but that may mean only that consciousness and free will are illusions. All these specific points reflect one underlying premise: for scientific materialists, the assumption of materialism is fundamental and not open to contradiction by evidence.

The disclaimer about religion reflected the same materialist way of thinking about the subject. For materialists, God "exists" as an object of subjective belief, just as Santa Claus and the tooth fairy exist in the imagination of children. Such beliefs are real for the people who hold them, and in moderation they may even be beneficial by providing comfort and encouraging moral behavior. Polite scientific materialists "respect" the religious beliefs of Christians and Jews just as they respect the equivalent beliefs of aboriginal tribes, but they do not take any of these beliefs seriously as truth claims. Only science, based on materialism, gives us objective knowledge that is valid for everyone. Outside of science there is only subjectivity, and in that realm any belief is subjectively valid "if it works for you."

The categories of materialist metaphysics explain why the science educators can insist in good faith that their teaching is not antireligious, while people who think God is the ultimate reality consider such disclaimers to be so much patronizing double-talk. The same understanding explains why Supreme Court justices can think the way to be "neutral" about the Bible is either not to teach it at all (while the public schools do teach everything the authorities think children need to know) or to teach it as "literature" (in other words, mythology). From a materialist standpoint, people who think the biblical accounts of the life of Jesus are historical are making what philosophers call a "category mistake," just as if they believed Dante's *Inferno* to be a scientific account of what lies under the earth. A man called Jesus may well have lived, and he may have taught some of the things ascribed to him, but he certainly did not walk on water or rise from the dead. On the other hand, it does no harm to believe in miracles if that gives you comfort—provided you do not carry things too far and place religious beliefs in the same category as scientific facts like evolution by natural selection.

It would be a mistake to describe the NABT fiasco as an incident in a conflict "between science and religion." The conflict needs to be more precisely defined. Many genuinely religious people, including clergy and professors at Christian seminaries, actually agree with the metaphysical categories of scientific materialism. Eugenie Scott likes to say that mainstream Christian ministers are among her most important allies in fighting creationism, and so they are. Relatively liberal Christians long ago made their peace with "evolution" and also with methodological materialism in science. By doing so they surrendered the academic world to the agnostics, but agnostics wisely leave a marginal corner for things like religious studies departments and for "Science and Religion" conferences, funded with gleanings from the harvest of government research money. The price for retaining that Christian reservation is to concede that science, based on materialism, trumps any other claims about "how things really are."

Even the Smith/Plantinga letter was framed within the categories of scientific materialism, which is why the NABT directors eventually found it persuasive. The letter did *not* say there is evidence that evolution was supervised, but rather that the question of supervision is a "theological question" that science can't address. It also pushed a reliable hot button by warning of the danger from "extremists in the religious right" if the barrier separating theological and scientific questions was not maintained.

I don't know who Smith and Plantinga (both personal friends of mine) had in mind as "extremists," but I do know what that term means to Eugenie Scott and the NABT directors. It means people who think God is objectively real as our true Creator and who think evolution by natural selection is merely a product of human subjectivity.

Epilogue: The National Academy's Uncertain Trumpet
In the aftermath of the NABT's embarrassment, while this book was in press, the prestigious National Academy of Sciences issued its long-awaited new guidebook for "Teaching About Evolution and the Nature of Science" (hereafter *Guidebook*). This publication takes so

cautious a stance that it suggests that the most authoritative voice of the scientific community is afraid to come to grips with any of the points in controversy.

A timid and evasive statement does not deserve an extended review. I'll merely set out some of the most important questions the National Academy might have been expected to answer and describe how it systematically fails to answer them.

1. Why is a new policy statement necessary at this time? If science education is currently failing to present the subject of evolution effectively, is it time to consider a new strategy?

The *Guidebook* at first identifies the problem as *unbelief.* Despite strenuous efforts by science educators over many decades, "fewer than one-half of American adults believe that humans evolved from earlier species." Most Americans think that some alternative to evolution should be presented in the schools or that evolution should be taught as theory rather than fact—by which they mean that it should be taught less dogmatically. The *Guidebook* responds to this perfectly intelligible point with a semantic quibble about what scientists mean by "theory."

When a strategy isn't working, the sensible response is to consider another approach. Why not try to communicate with the unbelievers rather than just overpower them? Instead of attacking the same old straw men, educators might identify the most responsible critics of naturalistic evolution and incorporate their objections into the curriculum. That might be a more effective means of persuasion, and it certainly would make the subject more interesting for students. The *Guidebook*'s authors opt instead for a head-in-the-sand approach that simply asserts that there is nothing to argue about. "There is no debate within the scientific community over whether evolution occurred," they assert, "and there is no evidence that evolution has not occurred." Since the *Guidebook*'s definition of evolution is so broad that every birth could be cited as a confirming example, the statement is both obviously true and utterly uninformative.

The educational opportunity that is thereby missed is illustrated by one of the few good points in the *Guidebook*. Recognizing that many otherwise excellent students will refuse to believe the official theory, the Academy commendably tells teachers that students should be graded only on their understanding of the theory, not on whether they believe it. The *Guidebook* illustrates the point in the words of a teacher who says of a creationist student, "I told her I wasn't going to grade her on her opinion of evolution but on her knowledge of the facts and concepts. She seemed satisfied with that and actually got an A in the class." But if even A students are sometimes skeptical of the grander claims of evolution, doesn't this suggest that there are reasonable grounds for doubt which a good science teacher ought to explore? It is not likely that bright students will be persuaded, or even interested, by a curriculum that effectively tells them that "some examples of evolutionary change can be cited, and therefore there are no controversial issues to discuss." Students who hear that line will figure out that they have to go to outside sources to learn the other side of the story.

2. What exactly is "evolution"? Specifically, are the science educators merely telling us that some change in living organisms has occurred, or do they claim to be able to provide a complete naturalistic account of the history of life, from the emergence of the first living organism up to and including the human mind?

The *Guidebook* defines *evolution* as "change in the hereditary characteristics of groups of organisms over the course of generations." This is what is commonly called microevolution, and no one, including biblical fundamentalists, denies its occurrence. The dispute concerns how new body plans and complex organs, including the human mind, came into existence. To the extent that the *Guidebook* addresses that issue at all, it does so only by persistently implying that the familiar examples of Darwinian microevolution illustrate a mechanism that, given sufficient time, is capable of creating all the immensely complex types of living organisms that now exist. The authors of the *Guidebook* know very well that there are many objections to this gigantic extrapolation from microevolution to a complete history of life, but they acknowledge the existence of scientific controversy only with the oblique statement "Some of the details of how evolution occurs are still being investigated." The mechanism of biological creation is not a "detail."

Although the *Guidebook* makes no explicit concessions, it appears from significant omissions that the National Academy is retreating from earlier claims that evolutionary scientists know (at least in principle) how the first life emerged from a prebiotic soup through chemical evolution. We hear nothing about the famous Miller-Urey experiment, which is continually presented in textbooks, museum exhibits and public television shows as if it demonstrated how chemical evolution can produce life. Instead of repeating the old claims about the origin of life, National Academy president Bruce Alberts conceded at a press conference that "that one is still up for grabs."

The *Guidebook* also has nothing to say about the origin of the cell, and it tells none of the usual "just-so stories" about the emergence of flight or vision. It does claim that whales descended from land-dwelling animals through three intermediate forms, and says, "It has been proposed that a small group of modern humans evolved in Africa 150,000 years ago and spread throughout the world." This kind of speculation does not address the main point. Even if we grant the evolutionary history as the *Guidebook* tells it, what was the source of the immensely complex genetic information required for the construction of a land mammal, a whale or a human being? For that matter, what was the source of the genetic information required to construct even a bacterial cell? The *Guidebook* makes no attempt to answer.

3. Does evolutionary theory have significant religious implications? Does natural selection shape human thought and behavior, as writers like Edward O. Wilson, Steven Pinker and Daniel Dennett say in their influential books, or is sociobiology a pseudoscience with little basis in scientific fact? When the authority of "science" is invoked to

buttress controversial theories about human nature, how can the
public distinguish genuine scientific knowledge from political ideol-
ogy?

The *Guidebook* gives students absolutely no help with questions
like these because of its authors' overwhelming concern to foster the
impression that scientists are agreed on all the important issues and
debate only the details. On religious issues, the *Guidebook* provides
only a bland evasion: "Within the Judeo-Christian religions, many
people believe that God works through the process of evolution." No
doubt they do. But does the National Academy consider the concept
of God-guided evolution to be scientifically acceptable, or is this
compromise position based on a misunderstanding that the scientific
elite tolerates only for political reasons?

The *Guidebook* evades the religious questions and ignores altogether the scientific questions relating to the implications of Darwinism for human behavior. Informed persons know that the Darwinists
are engaged in a civil war between "fundamentalists" and "pluralists,"
and that the hot-button issue in that war is the application of Darwinian
theory to human behavior. Recall that Richard Lewontin described the
sociobiology of Edward O. Wilson as resting "on the surface of a
quaking marsh of unsupported claims about the genetic determination
of everything from altruism to xenophobia" (see p. 70 in this book).
Yet Wilson and similar thinkers appear in the media and in college
curricula as scientific authorities, wielding the same Darwinian logic
that the National Academy urges the public to accept unquestioningly
when it applies to plants, animals and the human body. Should a good
education in evolution encourage young people to take sociobiology
at face value as scientific knowledge—or to be suspicious of it as one
of those fads, like Freudianism, that sometimes succeed for a while in
counterfeiting science without ever incorporating the requirement that
hypotheses be rigorously tested?

If the National Academy is serious about teaching evolution, as
opposed to just promoting a vague orthodoxy, it has to be prepared to

speak to the issues that students as future citizens most need to understand. One can sympathize with public school teachers who would rather encourage students to appreciate the virtues of scientific reasoning by presenting the topic in a less ideologically loaded context. But the overall message of the *Guidebook* is that "evolution" should be taught boldly as science, regardless of how many people object and regardless of who is offended. I actually believe in that message, and I am convinced that putting it into practice would be the best way of helping the scientific community to escape from the dead hand of Darwinian dogma. I only wish that the authors of the *Guidebook* had the courage of their own convictions.

Part 2

Essays on Books, Culture & Law

10

Engaging the
Third Culture

ON THE OPENING PAGE OF THE SELFISH GENE *RICHARD DAWKINS wrote, "We no longer [after Darwin] have to resort to superstition when faced with the deep problems: 'Is there a meaning to life? What are we for? What is man?' After posing the last of these questions, the eminent zoologist G. G. Simpson put it thus: 'The point I want to make now is that all attempts to answer that question before 1859 are worthless and that we will be better off if we ignore that completely.' "* *The Simpson/Dawkins philosophy animates John Brockman's vision of the Third Culture, as explained in his* The Third Culture: Beyond the Scientific Revolution *(Simon & Schuster, 1995); my review was first published in the inaugural issue of* Books & Culture, *September/October 1995.*

John Brockman is a Manhattan literary agent who represents scientists who write books for the general public. *The Third Culture* is a collection of taped interviews with twenty-three scientists, many of whom are clients of Brockman's agency, about their own scientific theories, what they think of each other, and above all their common ambition to be recognized as the true intellectual leaders of modern culture.

Among those represented in this volume are biologists Richard Dawkins, Stephen Jay Gould, George Williams, Brian Goodwin, Lynn Margulis and Niles Eldredge. In consciousness studies there is the fabled Marvin Minsky, along with Roger Schank, Daniel Dennett, Nicholas Humphrey and mathematician Roger Penrose. In the much-hyped field of complexity we find Murray Gell-Mann, Stuart Kauffman, Christopher Langton, J. Doyne Farmer and Daniel Hillis. Then there are the cosmologists: Martin Rees, Alan Guth, Lee Smolin and Paul Davis. Despite the absence of some superstars like Hawking, Crick and Weinberg, this is a splendid lineup.

Brockman's title refers to C. P. Snow's famous division between literary intellectuals and scientific intellectuals, groups that are said to inhabit different mental worlds. Snow predicted the emergence of a "third culture" of scientists capable of communicating with nonscientific intellectuals, but Brockman's authors intend to replace the literary intellectuals rather than cooperate with them. The opening sentence of this volume reads like an announcement of a hostile takeover: "The third culture consists of those scientists and other thinkers in the empirical world who, through their work and expository writing, are taking the place of the traditional intellectual in rendering visible the deeper meanings of our lives, redefining who and what we are."

Whether the scientific writers are displacing other kinds of intellectuals may be debatable, but there is no doubt that they are producing some very good books about science that have more intellectual substance than is suggested by the term *popularization*. Scientists have learned to write in plain English not only to communicate with the public but also to make sense to other scientists across disciplinary boundaries. Sometimes important new ideas are best discussed outside specialized research communities, where the main concern is to fill in the details of the current ruling theory. To pick the most famous example, Darwin's *On the Origin of Species* was written for the public and achieved a popular success. According to Daniel Hillis, "Many of the scientists who write popular books do so because there are certain kinds of ideas that have absolutely no way of getting published within the scientific community."

Brockman's scientists express their basic viewpoints succinctly in the interview format and provide candid comments on the value of each other's ideas. The effect is rather like sitting in on a crossdisciplinary seminar at the highest level. Reasonably well-informed readers will find these interviews particularly valuable for understanding the basic difference in outlook toward evolution between Gould and Dawkins, for example, or between Roger Penrose and the advocates of "strong AI." Strong AI (artificial intelligence) is the reductionist doctrine that was most colorfully stated by Marvin Minsky: the human mind is "a computer made of meat."

The interviews are illuminating, but they do not do very much to support Brockman's claim that scientists are pushing literary intellectuals aside and seizing control of the cultural high ground. Brockman's intellectual agenda is mainly limited to the well-worn naturalistic theme that "complex systems—whether organisms, brains, the biosphere, or the universe itself—were not constructed by design; all have evolved." We all know that scientists like to think that way and that any attempt to introduce God or the concept of design into a scientific discussion is likely to be repulsed with the emphatic assertion that "science" is defined by its commitment to methodological naturalism. That kind of intellectual rule-making does not tell us whether the scientists have actually *demonstrated* that design is absent from nature or whether they have merely assumed it.

Many Christian academics who want to reconcile their theism with contemporary scientific knowledge rely on making a distinction between "methodological" and "metaphysical" naturalism. Their point is that scientists are dismissing the possibility of a supernatural

Designer only from *science* and not from *reality*. A multitude of slogans can be called into service to support this distinction: the Bible is not a scientific textbook, science deals with the "how" and leaves the "why" to religion, Christians must at all costs avoid resorting to an embarrassing "God of the gaps," and so on. Together these slogans enforce a fundamental rule of contemporary intellectual life: theists who aspire to be tolerated in the academic world must accept the conclusions of "science" at face value, even if they suspect that the conclusions are influenced as much by philosophy as by empirical evidence, and they must do their theologizing from that bedrock foundation of neutral, unchallengeable "fact."

Yet the distinction between methodological and metaphysical naturalism seems pointless to the metaphysical naturalists who dominate contemporary science, and no hint of it appears in Brockman's interviews. To mainstream evolutionary scientists, the validity of naturalism as a worldview has been confirmed by science's success in providing all those unchallengeable facts that even theists dare not dispute. That is why Daniel Dennett can say that the strong AI position is "a very conservative extrapolation from what we know in the rest of science." If the unintelligent accumulation of random mutations by natural selection built all complex biological adaptations, then this Darwinian mechanism must also have built the human brain—and hence the mind and its consciousness. How then could the mind even conceivably be something human intelligence cannot duplicate?

The key assumption in that chain of reasoning is that natural selection had to have made the wonders of biology, because no scientifically acceptable alternative has been proposed. But for all the certainty with which brain scientists assume the vast creative power of natural selection, it seems that some biologists who actually study evolution have their doubts. Two of the biologists interviewed by Brockman (Brian Goodwin and Lynn Margulis) explicitly reject the neo-Darwinian model when it is extended beyond the modest finch-beak and peppered-moth-color examples where it finds its only empirical support. Other apparent critics of Darwinism such as Stuart Kauffman and Stephen Jay Gould seem to reject Darwinian orthodoxy at some times and accept it at others, but their ambiguities may be deliberate. According to the irrepressibly candid Daniel Hillis, there is "a strong school of thought in biology that one should never question Darwin in public," because "the religious right are always looking for any argument between evolutionists as support for their creationist theories." In such a delicate political situation we might expect skepticism about orthodox Darwinism to be expressed with less than perfect clarity.

From a naturalistic standpoint, arguments between evolutionary scientists about the adequacy of the neo-Darwinian mechanism in no way cast doubt on the philosophical claim that the possibility of design in biology has been conclusively refuted. From the point of view of those of us who want to know whether the nonexistence of a Designer is supported by evidence, as opposed to mere philosophical presupposition, the degree of support for a specific mechanism for generating adaptive complexity is all-important, and the evidentiary difficulties of the Darwinian scenario are highly significant.

I don't want to emphasize either the explicit or the implicit dissents from Darwinism, however, because the most revealing remark about Darwinism in *The Third Culture* comes from a Darwinist of unimpeachable authority, George Williams. Williams is much less visible to the public than Dawkins or Gould, but he is more authoritative in the profession than either. Along with John Maynard Smith and William Hamilton, he is at the summit of the inner circle of evolutionary biology, in a realm where Gould is regarded as a gadfly and Dawkins is something of a junior partner. Williams and Hamilton earned their preeminent status by pioneering the gene-centered Darwinism that Dawkins popularized with such success in *The Selfish Gene*.

In short, Williams is a top-flight authority and as orthodox a Darwinist as exists. Although his view of evolution is fundamentally the same as that of Dawkins, he criticizes Dawkins for describing the "gene selection" version of Darwinism as if the evolving Replicator were "a physical entity duplicating itself in a reproductive process"— that is, something like a section of DNA. According to Williams, the crucial object of selection in evolution is inherently nonmaterial:

> Evolutionary biologists have failed to realize that they work with two more or less incommensurable domains: that of information and that of matter. . . . These two domains can never be brought together in any kind of the sense usually implied by the term "reductionism." . . . The gene is a package of information, not an object. The pattern of base pairs in a DNA molecule specifies the gene. But the DNA molecule is the medium, it's not the message. Maintaining this distinction between the medium and the message is absolutely indispensable to clarity of thought about evolution.
>
> Just the fact that fifteen years ago I started using a computer may have had something to do with my ideas here. The constant process of transferring information from one physical medium to another and then being able to recover the same information in the original medium brings home the separability of information and matter. In biology, when you're talking about things like genes and genotypes and gene pools, you're talking about information, not physical objective reality. (p. 43)

Perhaps evolutionary biologists have avoided noticing that information and matter are fundamentally different things because that insight is fatal to the whole reductionist project in biology. If the message is truly not reducible to the medium, then trying to explain the creation of the information by a materialistic theory is simply a category mistake. One might as well try to explain the origin of a literary work by invoking the chemical laws that govern the combining of ink and paper, and then proposing speculative hypotheses about how those laws (with a boost from chance but without intelligence) might have generated meaningful sentences.

Neo-Darwinism is a theory of small-scale variation, not a theory of information creation. When Darwinists pay any attention to the information problem, they are satisfied to announce that new genetic information emerges mysteriously from a black box labeled "mutation." This fundamental gap in the theory seems to be tacitly recognized by some scientists, which is why there is so much interest in looking for new physical laws that might explain where the information comes from. But if a new kind of theory is needed, why are we so fervently urged to believe in the old one?

When members of a specialized research community aspire to be the interpreters of reality for everybody and announce their authority to redefine "who and what we are," they invite other intellectuals to examine their assumptions. Perhaps the heuristic assumptions the researchers have made for specific purposes are not suitable for other purposes and lead to error when carried too far. Doctrines that nobody in the scientific community dares to challenge may come under critical scrutiny from outsiders who do not have the professional mindset. I wonder whether the ambitious scientific intellectuals of the third culture are prepared for that to happen.

11

The Law & Politics
of Religious Freedom
A Revolution in the Making

THE SUPREME COURT'S DECISIONS ON RELIGIOUS FREEDOM REFLECT a mix of mutually contradictory notions. Law professor Steven D. Smith (Foreordained Failure: The Quest for a Constitutional Principle of Religious Freedom *[Oxford University Press, 1995]) and litigator William Bentley Ball* (Mere Creatures of the State? Education, Religion and the Court *[Crisis Books, 1995]) help us to see the intellectual and political forces that combined to create the mess. This consideration of their works was first published in* Books & Culture, *November/December 1995.*

Steven D. Smith is a University of Colorado law professor who deals with abstract intellectual issues; William Bentley Ball is a practicing lawyer who represents religious groups and individuals seeking freedom from dominance by government entities committed to secularism. Both men have valuable insights into a revolution that is currently under way in the interpretation of constitutional law regarding freedom of religion and religious establishment.

This revolution involves a changing understanding of what it means for the government to be "neutral" on religious questions, neither favoring nor opposing either particular religions or religion-in-general. For at least the last twenty-five years the dominant principle (occasionally ignored in practice) has been that neutrality means "no aid" to religion. The competing principle, which now seems to have five votes on the Supreme Court, is that neutrality means giving "equal treatment" to religious and nonreligious entities alike.

In a context where the government is giving substantial subsidies or benefits to nonreligious entities—such as secular educational institutions or student activities—the two principles have radically different consequences. Under the "no aid" principle it is unconstitutional (as an establishment of religion in violation of the First Amendment) for the government to broaden its subsidies to include religious schools or groups; under the "equal treatment" principle it is unconstitutional for the government to discriminate against those same religious entities by *not* broadening the subsidies.

A significant statement by the Supreme Court on the question is a 5-4 decision in *Rosenberger* v. *Rector* in June 1995. The case involved the Student Activities Fund (SAF) at the University of Virginia, which receives money from mandatory student fees and uses it to subsidize student activities, including the printing costs of student publications. Those publications are free to advocate all sorts of political and social causes, but the university invoked the "no aid" standard and refused reimbursement to a Christian student organization because its newspaper "primarily promotes or manifests a particular belief in or about a deity or an ultimate reality."

Four Supreme Court justices agreed with the university, citing cases holding that to provide public money for religious advocacy violates the no-establishment rule even if the same subsidy is given to all student publications regardless of their content. The majority of five justices cited other cases that followed the equal treatment rationale and held that by denying a subsidy only to student publications that advocate "religion," the university violated the religious students' right to freedom of expression.

In his wonderfully clarifying book Smith explains why it is impossible to resolve this clash of principles either on the basis of the language of the Constitution or on the basis of historical evidence that the First Amendment was intended to prefer one principle or the other. Sometimes the "intent of the Framers" is ambiguous, but Smith explains that in the case of the First Amendment's religion clauses the original intention is perfectly clear, and the post-World War II Supreme Court has simply chosen to disregard it.

The whole point of the First Amendment religion clauses was to deny jurisdiction over religious questions to the federal government and leave such matters to the discretion of individual states. Thus the amendment was drafted to say that *"Congress* shall make no law respecting an establishment of religion." The national legislature was not to meddle in matters of religious establishment at all, whether by instituting such an establishment at the national level or by interfering with the religious establishments (e.g., state payment of clergy stipends) that existed in some states.

The question of the proper relationship of government and religion was controversial in the late eighteenth century just as it is now, and there is no way of knowing how it would have been resolved if the Framers had decided to tackle it. That is why Smith's title declares the quest for a national constitutional principle of religious freedom to be a "foreordained failure": the point of the religion clauses was precisely to prevent the formation of such a principle by leaving the matter to the states. When the twentieth-century Supreme Court declared that the religion clauses were "incorporated" into the Fourteenth Amendment's due process clause and hence were applicable to state and local governments, the Court effectively reversed the intent of the Framers and declared itself to be the national religious lawmaker that the First Amendment was expressly intended to forbid.

If the intent of the Framers were to be our guide, then what is clearly unconstitutional is practically everything the Supreme Court has done in this area since 1947. Yet it is unlikely that the Court will ever repudiate its usurpation of authority, especially since the public has grown accustomed to thinking of religious questions as matters to be resolved at the national level.

The mere fact that five justices currently support the equal treatment principle does not imply a well-established new outlook, since the next appointment might shift the balance back the other way. (The two Clinton appointees, Justices Steven Breyer and Ruth Bader Ginsburg, joined Justice David Souter's dissent in *Rosenberger*.) A possibly more enduring change in the ideological climate, however, is that the philosophical bottom has dropped out of the notion that between theism and agnosticism there is a secular rationality that is truly "neutral." Just about everybody in academia now understands that controversial and politically loaded value choices usually lie concealed behind the purportedly neutral rationalizing of power holders such as Supreme Court justices.

Is a school district neutral on religious questions when it leaves all mention of God and the Bible out of the curriculum—while purporting to teach students just about everything they need to know, from "values clarification" to how to use a condom? Smith quotes University of Utah Law School professor Michael McConnell to the contrary: "If the public school day and all its teaching is strictly secular, the child is likely to learn the lesson that religion is irrelevant to the significant things of the world, or at least that the spiritual realm is radically distinct and separate from the temporal." Protestants are at last realizing what Catholics understood all along: the notion that a religion-free secular knowledge is all we really need is anything but neutral on religious questions.

William Bentley Ball looks at the religious liberty issue from the perspective of a Pennsylvania lawyer who has represented religious groups and individuals in a variety of significant cases. These include his victory in *Wisconsin* v. *Yoder* (1972), in which the Supreme Court upheld the right of the Amish to keep their children out of public high school, and especially his stunning loss in *Lemon* v. *Kurtzman,* the 1971 Supreme Court decision that most firmly entrenched the "no aid to religion" principle into constitutional law. As a litigator, Ball believes that the voluminous writings of theoreticians in the law schools need to be balanced by some input "from below," to illuminate how constitutional rules feel to the people who are most affected by them. Speaking as one of those theoreticians, I thoroughly agree with him.

The issue in *Lemon* v. *Kurtzman* was whether the state of Pennsylvania could subsidize the purely academic side of education in religious schools, such as teachers' salaries and other expenses relating to math, foreign languages and so on—and exclude religious instruction, which would continue to be financed entirely from private sources. The rationale for allowing the subsidy was that education in these subjects was not substantially different in religious and secular schools, that religious schools were providing a public benefit by educating pupils who would otherwise have to be educated entirely at public expense, and that parents exercising their right to choose religious schools for their children are also taxpayers and ought to get some benefit from their taxes.

Ball was at first confident of victory, because the Supreme Court had previously upheld laws providing bus transportation and textbooks to religious schools, thus indicating that the purely secular side of their activities could receive public money. The Pennsylvania program led to a pitched legal battle, however, in which the religious schools were on the defensive for three reasons. First, almost all the schools that would benefit were Catholic schools; second, the public school lobby vigorously opposed the subsidy for economic reasons; and third, the subsidy was also opposed by both the Pennsylvania Council of (Protestant) Churches and the largest Jewish organizations in the state.

Believers in God were thus thoroughly divided, and many influential people saw no good reason for Catholics to be so determined to avoid a public school system that seemed to satisfy everybody else. In the circumstances the secularists persuasively characterized the measure as a sop to the political power of the Catholic Church. The Supreme Court justices in turn regarded that church with undisguised suspicion. The Court held that the subsidy was unconstitutional because the "secular" teaching in religious schools could not realistically be separated from religious indoctrination without a pervasive state supervision that would itself entangle the state in controversial religious affairs.

Ball's analysis suggests that the Supreme Court might well have approved a similar measure in a different context, and no doubt he is right. One needs only to look at the contrasting decision in *Wisconsin* v. *Yoder*, in which the same Court granted Amish families an extraordinary exemption from compulsory school attendance laws, to see that the justices had no absolute objection to conferring a protective benefit on a religious group regarded by everyone as appealing rather than threatening. On the other hand, the "balanced treatment for creation-science" legislation never had a chance of success in the Supreme Court, because the fundamentalists who were thought to be the only people opposed to "evolution" were as politically isolated as the Catholics in *Lemon* v. *Kurtzman*.

Christian and Jewish theists can draw at least two important lessons from the sad story related in these two books. First, we should never be impressed by arguments that the Constitution absolutely forbids some sensible measure that treats religious and secular interests fairly. What doomed the religious school subsidy was not a document locked up in the National Archives building but the dominant attitude toward Catholic schools at the time among the groups that Supreme Court justices take most seriously. Second, the people of God need to learn to unite on first principles before we start arguing over what follows. The Pledge of Allegiance that we all recite tells us that this is one nation under God. If that language rings hollow today, it is not primarily the fault of the agnostics but of the people who know God but have preferred to fight over what divides them rather than to unite over what they have in common.

12

How the Universities Were Lost

THIS ESSAY DESCRIBES TWO EXCELLENT BOOKS (GEORGE MARSDEN'S The Soul of the American University: From Protestant Establishment to Established Non-belief *[Oxford University Press, 1994] and Douglas Sloan's* Faith and Knowledge: Mainline Protestantism and American Higher Education *[Westminster John Knox, 1994]) that together show how and why Christians forfeited their standing in the intellectual world. "They" didn't take it from us; "we" gave it away. My review was first published in* First Things, *no. 51 (March 1995).*

S ociologist Peter Berger famously observed that if India is the most religious nation in the world and Sweden the least religious, then the United States of America is a nation of Indians ruled by Swedes. If you want to know how "Swedish" ruling power was established and consolidated in the United States and why the Swedes have been able to defeat easily all challenges from the Indians, you can do no better than to read the two books under review here.

The agnostics rule America, quite regardless of the popular piety to which politicians pay lip service, because their metaphysics (i.e., scientific naturalism) rules the universities, and the universities control the social definition of knowledge.

George Marsden's book tells the complete story in greater detail than many readers will require, since basically the same thing happened at many different universities. Douglas Sloan concentrates on a single episode. I enthusiastically recommend both books in their entirety, but readers with limited time might consider reading in Marsden the triple prologue, the chapters on the Universities of Michigan and Chicago, the final chapter and the "Concluding Unscientific Postscript." In Sloan's book chapter four is crucial; it is aptly titled "The Theologians and the Two-Realm Theory of Truth." Once you get the basic idea, the rest is a matter of (very interesting) detail.

Marsden's subtitle says it all: the story of the modern American university is that of a long march from Protestant establishment to established nonbelief—or rather, to established belief in scientific naturalism. In the nineteenth century things like compulsory chapel services were common even in state universities, and Protestant Christianity was just about everywhere (except in Catholic universities, of course) considered to be the governing ideology of higher education. When William F. Buckley's 1951 book *God and Man at Yale* charged Yale University's faculty with undermining the university's traditional commitment to Christianity (and conservative values), university authorities responded indignantly and with apparent sincerity that Yale remained Christian "in a broad sense." Today if the president and trustees of Yale were to proclaim that Yale is in *any* sense a Christian university, they would be met with a firestorm of protest—or howls of laughter.

At Duke University, famous today for basketball championships and postmodernist literary theory, a plaque at the center of the campus states, "The aims of Duke University are to assert a faith in the eternal union of knowledge and religion set forth in the teachings and character of Jesus Christ, the son of God." That was what Duke officially stood for at its initial endowment in 1924, and many other universities would then have articulated their mission in similar terms. When Duke formulated a new mission statement in 1988, however, its aims had become entirely secular in character, stressing only values like "the spirit of free inquiry" and the promotion of "diversity and mutual tolerance." The university's previous Christian identity was relegated to history with a statement that "Duke cherishes its historic ties with the United Methodist Church and the religious faith of its founders, while remaining nonsectarian." The new mission statement made clear, in Marsden's words, that "Christianity as such is peripheral to the main business of the university" today.

Yet the 1988 statement was not exactly a repudiation of Duke's original starting point in progressive Methodism. It would be more accurate to say that the agnostic university of the 1980s was the logical culmination of a trend that was present in liberal Christianity from the beginning. To hold on to its ruling position in a pluralistic democracy, the Protestant establishment had to become "inclusive," which meant that it had to suppress the distinctively Christian and supernatural elements in biblical theology. Becoming nonsectarian meant at first only cutting formal denominational ties. Because Catholics, Jews and eventually atheists had to be brought into the tent, however, "nonsectarian" eventually came to mean "wholly secular."

Christianity "in a broad sense" was merely a spiritualized version of Enlightenment rationalism, in which natural science claimed sole authority to describe reality, progress claimed the role of God, and social reform claimed the status of salvation. By the early twentieth century, the "union of knowledge and religion" of Duke's plaque looked more like a watered-down version of Marxism (with history marching moderately forward into a social utopia) than anything Calvin or Luther would have recognized as Protestant Christianity. When the tumult of the 1960s struck, the universities shed their Christian veneer without noticing that they were missing anything. The title of Marsden's final chapter describes the end result: "Liberal

Protestantism Without Protestantism"—and especially without God.

The irresistible secularization of the universities illustrates the principle that an establishment of religion sometimes does more damage to the religion being established—by wedding it to the culture—than it does to the unbelievers and dissenters who are supposedly the victims of discrimination. There was no fight to the finish between Christian theists and secularizers. The story Marsden tells is one of Christians in positions of formal power yielding willingly to each stage in the advance of secularism. Obviously there was some element in the situation that made Christian (or theistic) resistance to secularization ineffective. What was it?

Readers can get the answer from either book, but I think Sloan makes the essential point particularly clear. The liberal Protestant establishment wanted to combine a scientific picture of reality with the highest religious and ethical ideas. What was overlooked, in Sloan's words, was the "fact that nineteenth-century science, when viewed within its prevailing interpretive framework, was fundamentally at odds with religious and ethical ideas of any kind. This was clear to all who had eyes to see and who were able to look without flinching." All the strenuous efforts to keep Christian theism alive in a university dominated by scientific naturalism only amounted to so much flinching before the disquieting implications of the scientific outlook.

The crucial issue in the universities, according to Sloan, is the faith-knowledge dichotomy. From a scientific point of view, "knowledge" is inherently empirical, coming from sense experience and scientific investigation. This is the legacy of positivism, a philosophy that achieved its culminating triumph in the Darwinian theory of evolution. In modern universities professors take for granted that the universe began with something like particles in mindless motion governed by impersonal laws and that everything that has appeared since is the product of a purely naturalistic process of physical, chemical and biological evolution. "Everything that has appeared since" includes things like human religious and ethical beliefs, which are themselves presumed to be products of things like brain chemistry and natural selection.

The worldview of scientific naturalism preserves a place for religious beliefs: a place, that is, among the things to be explained by science. The Christian religion thus enters the university with a status precisely equal to that of other comparable religious systems—say, the Aztec system of human sacrifice. Any individual, even a person of eminence in science, can make a personal choice to "be religious." Such choices are made on the basis of "faith," meaning subjective preference. A problem arises only if the Aztecs or the Christians claim access to knowledge. If they do that, they are claiming that their own beliefs are normative for unbelievers. Only scientists can claim that kind of authority, because what is endorsed by the scientific community constitutes knowledge, not belief. That is why Darwinian evolution can be taught in the schools as fact, however strongly parents or students object, whereas a simple prayer acknowledging God as our Creator is deemed unacceptable—because somebody might object.

Sloan's basic thesis is that any person who wishes to assert the viability of Christianity in the modern university must come to grips with the faith-knowledge problem. Of course a scholar can be a Christian as a matter of faith, regardless of his or her attitude toward scientific knowledge, just as anyone can choose on the basis of faith to be a Muslim or a Rastafarian or a radical feminist. But prestige in the university goes only to those commitments that are seen as capable of generating knowledge. Once science has provided knowledge (Copernican astronomy, Newtonian physics, Darwinian evolution, quantum indeterminacy), various subjective ideologies can fit their belief systems into the framework of that knowledge. Intelligent people are unimpressed with this trimming and prefer to give their allegiance to the metaphysical system that provides the knowledge to which other systems must do their best to conform.

All efforts to assert Christianity in the university ended in futility because of Christians' inability or unwillingness to challenge naturalism's monopoly over the production of knowledge. Sloan's chosen example is the failed "theological renaissance" of the 1940s and 1950s that featured Reinhold Niebuhr, his less famous but equally esteemed brother H. Richard Niebuhr, and Paul Tillich. These "theological reformers" understood that a purely scientific knowledge is inadequate and denies the reality of personhood and meaning. Hence they tried to assert that factual knowledge has to be interpreted in the light of principles that transcend what is knowable scientifically. At the same time, they wanted to speak in a language intelligible to modernists, which meant accepting all the achievements of science and of critical analysis of the Scriptures. Thus the theological reformers "never tired of heaping ridicule on all religious persons who rejected Darwinism, lumping them all together as unreconstructed fundamentalists—even though Darwinism represented in fact the extension of the mechanical philosophy to everything, including persons and spirit, and was a direct challenge to any theological conception of a meaning in history that transcends history."

As Sloan sums up the results:

> In the end the theologians pulled back from affirming unambiguously the real possibility of knowledge of God and of the spiritual world. They again and again resisted seeking or talking about knowledge of God for fear of the danger of applying objectifying and manipulative modes of thought where they did not belong. At the same time, however, they wanted to affirm fully and without question, lest they be thought religious fundamentalists, the same objective, analytic modes of modern science and historical analysis in every other domain besides faith. The result was a split that forced the theological reformers back onto faith presuppositions whenever they spoke about religion, and onto an increasing reliance on naturalistic approaches to the sensible world whenever they wanted to speak about ethics, science, or knowledge in general.

In short, whenever the chips were down, Tillich and the Niebuhrs effectively conceded the high ground to the scientific naturalists. It should come as no surprise that a theological movement based on a tacit acceptance of naturalistic metaphysics was blown away like so much dust when the passionate political movements of the 1960s emerged. Exactly the same fate awaits all religious revivals, however impressive they may seem for a time, if they lack the nerve or the intellectual resources to challenge the cultural assumption that knowledge comes only from a science based on naturalism.

13

Wundergadfly

ALTHOUGH WE WERE BOTH PROFESSORS AT BERKELEY DURING THE 1960s and 1970s, I never met Paul Feyerabend. Before about 1980 I was a conventional law school careerist and had no idea that our intellectual interests would converge dramatically. I reviewed his autobiography, Killing Time *(University of Chicago Press, 1995), in the March/April 1995 issue of* Books & Culture.

If I had to describe Paul Feyerabend in two words, *brilliant* and *irresponsible* are the two that would immediately come to mind. Both qualities are on display in his engrossing autobiography, completed just before the author's death from a brain tumor in 1994.

Let's start with the brilliance. As a young man in Hitler's Austria, Feyerabend trained to be an opera singer; he had a promising future in that profession when he was drafted into the army in 1942. He rose through the ranks to become an officer and ended up commanding a battalion in the last stages of the disastrous retreat on the Russian front.

His physical courage earned him the Iron Cross and wounds so severe that for the rest of his life he was on crutches, in continual pain and sexually impotent. Despite these incapacities he was fabulously successful as a scholar, a lecturer, a connoisseur, a lover and a raconteur.

After the war Feyerabend studied physics and then more or less drifted into philosophy of science. He started as a protégé of Karl Popper but soon carved out his own position and became notorious as the leading voice for "epistemological anarchism," the precursor of what today we call postmodernism. In his most famous book, *Against Method* (1975), Feyerabend denied that there is any single form of reasoning that can be labeled "the scientific method" and asserted brazenly that the basic rule in science is "anything goes." Many scientists were not amused.

Although his irreverence outraged conventional scientists and philosophers, Feyerabend became and remained an academic superstar. He taught at Berkeley for most of his career but was constantly wooed by other prestigious universities and accepted or rejected their offers according to his mood of the moment. That he never got into serious trouble either with the Nazis or with the liberal academic elite indicates that he knew how to be provocative without saying anything unforgivable, and indeed Feyerabend admitted that his thinking was made up of "a rather unstable combination of contrariness and a tendency to conform."

That brings me to the irresponsibility. Feyerabend was the kind of professor who sometimes failed to show up for classes, who didn't want an office because he didn't want to hold regular office hours, and who was always flying off to give lectures somewhere else. At Berkeley he ostentatiously took the side of the student radicals, apparently for no deeper reason than that he enjoyed all the hell-raising. He was profligate in love most of his life, until his fourth wife turned him into a devoted husband who hoped to become, with medical assistance, a father.

As a philosopher Feyerabend was particularly concerned with the tension between truth and freedom. Once we have found some final truth, something that is true beyond question, must we give up our freedom to doubt? Jesus claimed to be the only way to the Father and also said, "You shall know the truth, and the truth shall make you free." This is a scandal to modernists, for whom the idea of absolute truth implies oppression, not freedom.

C. S. Lewis memorably caricatured the modernist mentality in *The Great Divorce*, in the form of a theologian who refuses to enter heaven unless the celestial powers guarantee that he will find there an atmosphere of free inquiry. The ministering spirit responds that, on the contrary, "I will bring you to the land not of questions but of answers, and you shall see God." The theologian spurns this "ready-made truth which puts an end to intellectual activity" and opts for hell—where, as in a deconstructionist English department, the complete absence of truth allows an unlimited scope for interpretation.

But modernism has its own exclusive road to truth. As a young scholar Feyerabend was a positivist who dismissed all statements about God from serious consideration by proclaiming that "the idea of a divine being simply had no scientific foundation." He explained that "science is the basis of knowledge; science is empirical; nonempirical enterprises are either logic [as in mathematics] or nonsense." People who say things like that have seen a God whose name is Science, and a very jealous God it is.

The later Feyerabend came to see science not as the only road to truth but merely as one of the ways of interpreting reality that a pluralistic society ought to include. In a famous 1974 lecture, "How to Defend Society Against Science," he argued that we should regard *all* ideologies, science included, "like fairytales which have lots of interesting things to say but with also contain wicked lies." He argued that although science was a liberating influence in the seventeenth and eighteenth centuries, in contemporary times it had become another stifling orthodoxy.

In education, he charged, scientific "facts" are now taught just as religious "facts" were taught a century earlier, with little attempt to stimulate students' critical faculties. At the professional level, "most scientists today are devoid of ideas, full of fear, intent on reproducing some paltry result so that they can add to the flood of inane papers that now constitutes 'scientific progress' in many areas." Nonetheless, "the judgment of the scientist is received with the same reverence as the judgment of bishops and cardinals was accepted not too long ago." Even theologians pursue a project of "demythologization" on the assumption that in any clash between science and religion, religion must always be in the wrong.

Feyerabend anticipated the obvious retort to this indictment, which is that science has earned its preeminence not by suppressing dissent but by discovering truth. "For once we have discovered the truth—what else can we do but follow it?" He responded that "a truth that reigns without checks and balances is a tyrant that must be overthrown. . . . My criticism of modern science is that it inhibits freedom of thought. If the reason is that it has found the truth and now follows it, then I would say that there are better things than finding and then following such a monster."

The last sentence echoes C. S. Lewis's damned theologian, who by renouncing truth achieved not freedom but absurdity. Without the goal of truth at the end of the process, freedom of thought is an exercise in futility, like a treasure hunt without the treasure. In reading Feyerabend we always have to discount the overstatements, but when we do so we often find that his real point was perfectly sensible. In this case Feyerabend was not renouncing the search for truth or implying that we can preserve freedom by repealing the law of gravity. He was proposing that we should encourage competition rather than monopoly in epistemology, just as we do in government (by separation of powers), in religion (by the prohibition of a religious establishment) and in the economy (by antitrust laws).

Up to a point this is orthodox. Ask any scientist why science is reliable, and he or she will cite the checks and balances, such as repeatable experiments, peer review, unfettered debate and the fierce competition for prizes. Whether these mechanisms always operate as advertised, especially when political or financial interests are involved and funding is centralized, is an important question that I will not attempt to address here. The broader point is that even under ideal circumstances, scientific debates occur only within the profession, which means among those who share the professional mindset.

In this respect science is like the medieval church, which permitted theological debates among scholars (in Latin) but expected the laity to leave judgments about such matters to the clergy. Feyerabend wanted to break the clerical monopoly by making room in scientific debates for persons who know about science "without being taken in by the ideology of science." That ideology is roughly the position the young Feyerabend advocated and then outgrew, whether it is called positivism, empiricism, naturalism, materialism or scientific atheism. Is reality truly limited to the things scientists can study, or should science itself take account of a reality outside the ken of science? Only outsiders can raise questions like that. If insiders tried to raise them, they wouldn't be insiders for long.

The need for outside perspectives in science has grown in the twenty-one years since Feyerabend delivered his lecture. For example, in *The New York Review of Books* for November 30, 1995, the eminent Darwinist John Maynard Smith brought into the open the bitter schism in evolutionary biology over issues like "gradualism" and "adaptationism." In the same essay Maynard Smith offhandedly pronounced ideological judgments as if they were findings of biology. "We see humans as the joint products of their genes and their memes—indeed, what else could they possibly be?" From this philosophy dressed up as biology Maynard Smith derives moral relativism, because "if a person is simply the product of his or her genetic makeup and environmental history, including all the ideas ['memes'] that he or she has assimilated, there is simply no source whence absolute morality could come." Other evolutionary biologists frequently say this sort of thing because they have lost sight of the difference between ideology and science, if they ever recognized a difference.

Evolutionary biology now consists of at least two factions that disagree fundamentally over how evolution is supposed to have occurred but share a determination to exclude the "creationists," meaning all those millions of people who think that a Creator may have had something to do with the history of life. As they learn that what is at stake is not the Genesis chronology but the very idea that a source of absolute morality could conceivably exist, more and more people are going to insist on a right to participate in the arguments that divide the ideologists of evolutionary biology.

I would like to think that Paul Feyerabend, wherever he may be, is looking on and enjoying the fun.

14

Gideon's Uncertain
Trumpet

THIS ESSAY IS NOT A BOOK REVIEW IN THE ORDINARY SENSE. IT WAS originally published as the introduction to the Legal Classics edition (1994) of Anthony Lewis's Gideon's Trumpet *(1964). As you will see, a lot happened in the intervening thirty years. I have removed all the usual cumbersome legal citations because they would be distracting for most readers. Any law student can tell you how to find all the cases.*

Anthony Lewis is known to newspaper readers today as an angry man, a liberal *New York Times* Cassandra who deplores the folly of voters who keep on electing presidents like Nixon, Reagan and Bush. *Gideon's Trumpet* was the product of an earlier and mellower Lewis who was at home in the America of 1964. In those days the liberal reforming tradition had retaken possession of the White House following the mildly conservative Eisenhower interregnum. Richard Nixon had apparently self-destructed in a spectacular press conference after losing an election for governor of California. The martyred John Kennedy had been succeeded by the far more effective Lyndon Johnson, who was on his way to a landslide victory over Barry Goldwater. Democrats dominated the Congress, and the judicial branch had become the spearhead of liberal reform. Ironically, the Republican president Eisenhower had unwittingly done more than anyone to solidify a liberal activist judiciary, by appointing Earl Warren and William Brennan to the Supreme Court.

Liberals thus enjoyed control of all three branches of the federal government. They also had virtually undisputed possession of the moral high ground. Conservatives were reeling under the moral burden left by Joe McCarthy and Jim Crow. When Barry Goldwater called for law and order, the media took for granted that he was making a coded appeal to racial prejudice. The Vietnam debacle, the New Left and urban riots were in the future. Liberals had no idea that they would soon have to deal with such divisive issues as radical feminism and gay liberation. The egg of rationalistic skepticism had been laid in the universities, but its nihilistic offspring had not yet begun to deconstruct the shell. The first half of the sixties was a good time to be a liberal.

Gideon's Trumpet celebrates the reforming spirit of that era and shows us a Supreme Court at the height of its unity and prestige in the aftermath of that great statement of moral purpose known as *Brown* v. *Board of Education* (the famous school desegregation case of 1954). The book tells three good stories. First is the story of Clarence Earl Gideon, a small-time gambler, petty thief and five-time loser. Gideon achieved the convict's equivalent of winning the Irish Sweepstakes. His hand-lettered petition to the Supreme Court led to a landmark legal decision and to freedom for himself. The Supreme Court ruled in 1963 that Gideon (and every other felony defendant in the United States) was entitled to a free defense lawyer if he could not afford to pay. Gideon got a new trial, at which the jury found him not guilty, and so he left prison and entered legal history as a folk hero.

The second story is about the Supreme Court and how it is possible for an indigent state prisoner like Gideon to obtain justice from the nation's highest court. Lewis's narrative follows Gideon's petition from the prison to the Supreme Court Clerk's office, from there to the law clerks who evaluated it for the justices, and from the law clerks to the justices who decided in conference to grant a hearing and appoint counsel to argue for Gideon. The Supreme Court bureaucracy has grown in the intervening three decades, and procedures have changed superficially, but in principle the same story could be told today. Handwritten petitions still arrive at the Court from indigent prisoners, and they are still evaluated by able young law clerks eager to find an exciting issue. Every so often the Court selects one for oral argument and decision.

The lawyer appointed by the Supreme Court to argue for Gideon was the renowned Abe Fortas, head of a powerful Washington law firm and confidant of Lyndon Johnson. (Fortas later became a Supreme Court justice himself but was forced to resign due to financial improprieties.) Like any busy senior law partner, Fortas relied on junior colleagues for research and writing. One of his assistants was a Yale law student named John Hart Ely, who subsequently achieved fame as a constitutional law scholar and dean of the Stanford Law School. Fortas and company argued Gideon's case so well (against an overmatched lawyer representing the state of Florida) that they received the votes of all nine members of the Court. The underdog won, as underdogs tend to do when they are represented by the most powerful voices in the legal profession.

The third story is about the Constitution and how its meaning changes to conform to new expectations. This story begins with the infamous "Scottsboro boys" case of the 1930s, in which a group of young black drifters were convicted of interracial rape in Alabama. Community feelings ran so high against the defendants that nobody wanted to defend them, and so the judge ceremonially appointed all the local lawyers to act as defense counsel. Having everybody for a lawyer is about the same as having nobody for a lawyer. For this and other reasons, the trial looked a lot like a lynching.

In one of the resulting appeals (*Powell* v. *Alabama,* 1932) the Supreme Court held that indigent defendants faced with the death penalty have a constitutional right to the assistance of appointed counsel. Some lawyers thought the right applied to all criminal cases punishable by imprisonment, but in 1942 there was a retrenchment. In *Betts* v. *Brady* the Supreme Court held that a Maryland felony trial was fairly conducted even though the trial judge had refused to appoint a lawyer for the indigent defendant. The case was routine and the judge was fair, said the majority, and that was all "due process of law" required.

If the trial had been held in a federal court, the Sixth Amendment would have commanded appointment of counsel. The specific requirements of the Sixth Amendment apply only to the federal government, however, except to the extent that the Supreme Court has held them to be "incorporated" into the due process clause of the Fourteenth Amendment, which governs state procedures. State courts, said the *Betts* majority, had to supply counsel to indigents only where "special circumstances" render the case particularly difficult or complex.

The rule of *Betts* v. *Brady* proved unsatisfactory, because it is a lawyer's job to find and articulate the circumstances that make a case special. To appoint counsel only where those circumstances are already apparent is to guarantee that "special" cases will be misperceived as "routine." Hence Justice Hugo Black, a former Ku Klux Klan member whose better nature had internalized the logic of *Powell* v. *Alabama,* understood by visceral instinct that *Betts* v. *Brady* was wrongly decided. Other justices took longer to come to the same point of view, but experience in trying to apply the "special circumstances" rule gradually brought them around. By the time of Gideon's case a majority had read the handwriting on the wall. Knowledgeable observers expected the Court to overrule *Betts* v. *Brady* sooner or later, and probably sooner.

Matters had even progressed to the point where many prosecutors and trial judges were convinced that every felony defendant should be entitled to a lawyer. All the states outside the Deep South were routinely appointing counsel for indigents. When the Florida attorney general circulated a letter asking other states to support his argument in favor of retaining the *Betts* doctrine, the results were worse than disappointing. A young man named Walter Mondale, then attorney general of Minnesota but headed for greater things, replied that he favored a right to counsel for all defendants. One thing led to another, and eventually the chief law enforcement officers of twenty-three states endorsed a "friend of the court" brief that took Gideon's side of the case. Fortas doubtless made a fine argument, but it was Mondale and his colleagues who ensured the unanimous victory.

The aftermath of the Supreme Court decision takes us back to the personal history of Clarence Earl Gideon. Gideon was more successful as a writ-writing prison lawyer than he ever was at anything else, and his triumph encouraged him to develop some heady notions about the law and about his own abilities. For a time he seemed set upon reducing the great constitutional drama to farce by demanding the right to represent himself at the retrial. The state judge adamantly refused to allow such an absurdity, and Gideon finally agreed to accept a local lawyer named Turner. The judge quickly appointed Turner as defense counsel and sternly warned Gideon not to interfere with his handling of the case.

Lewis's account of the retrial shows how much difference it can make to have a competent defense lawyer cross-examining the prosecution's witnesses. Turner spent three full days interviewing prospective witnesses and investigating the facts. The theory he settled on was that the crime had actually been committed by the prosecution's witness Henry Cook, who had fingered Gideon to clear himself. Cook was an unreliable witness and apparently the kind of person who might have done something of the sort, and so the jury rightly found that Gideon's guilt was not proved beyond a reasonable doubt.

Gideon went free, and so did a few thousand other prisoners from Florida and neighboring states who had been convicted of felonies without the assistance of counsel. No great crime wave resulted. Besides its immediate effect, the *Gideon* decision focused attention on what might be called the bureaucratic problems of providing defense counsel under contemporary conditions. In a rural society with little crime, lawyers can share the occasional burden of representing indigents on a charitable basis. A mobile urban society, however, generates crime of a different quantity and quality. Provision of legal services has to be regularized, and fair compensation must be provided. Lewis recounts briefly how bar groups and governmental bodies began to take up the challenge of financing and administering the right to counsel after they were awakened by the sound of Gideon's trumpet.

Some readers will want to know how the constitutional right to counsel developed after the period covered by Lewis's account. To address that subject, I have to begin by explaining a decision Lewis surprisingly fails to discuss; the "other" right to counsel case which the Supreme Court decided on the same date as *Gideon* v. *Wainwright.* This was *Douglas* v. *California,* and it involved the right to counsel on appeal after conviction. California provided free counsel to all indigents at trial, but for appeals it followed the approach of *Betts* v. *Brady.* The appellate court appointed counsel for the convicted person only after making a preliminary determination from the record that such appointment "would be of advantage to the defendant or helpful to the appellate court."

What made *Douglas* v. *California* tricky was that the Constitution does not mention a right to appeal in criminal cases. On the contrary, the Supreme Court had said in an 1894 decision that the states have no constitutional obligation to permit appeals from criminal convictions. If the Constitution does not compel the states to provide an appeal at all, it does not compel them to provide a right to counsel on appeal. The holding in Gideon's case—that the Sixth Amendment right to counsel is applicable to the states through the Fourteenth

Amendment's due process clause—was no help to Douglas.

On the other hand, the Supreme Court had said in 1956 *(Griffin* v. *Illinois)* that a state that does allow appeals (as all states in fact do) must not effectively bar poor defendants from appealing by requiring them to pay for a transcript of the trial. (Without a transcript the appellate court does not know what happened at the trial and hence cannot identify any errors.) This decision was based on the Fourteenth Amendment's "equal protection" clause. The Supreme Court invoked the *Griffin* decision and the equal protection clause in *Douglas* to hold that states must appoint counsel for indigents in all felony criminal appeals, because "there can be no equal justice where the kind of appeal a man enjoys depends on the amount of money he has."

Many legal scholars thought *Douglas* was a more important decision than *Gideon,* because the equality principle is potentially more far-reaching than the due process principle. It is one thing to give a defendant a lawyer; it is quite another thing to provide a defense as good as a millionaire might purchase. When the liberal early sixties became the radical late sixties, the possibility that the Constitution might require some prodigious effort at equalization of wealth—and not only in the criminal process—began to be taken seriously. At the very least, the equality principle provided social critics with a powerful rhetorical weapon, since wealth unquestionably makes a difference in the criminal justice system.

Here, as briefly and nontechnically as the subject permits, is what all this came to mean after the dust settled. *Gideon*'s right to counsel was extended from felonies (punishable with a year or more of prison time) to misdemeanors where any jail time is imposed upon conviction. The equality-based right to counsel on appeal applies only to the first stage of the appellate process, usually a state or federal intermediate court of appeal. Defendants who want to seek further review in a state supreme court or the United States Supreme Court have no absolute right to legal assistance, although of course they can make use of the brief counsel filed for them in the initial appeal. The discretionary approach of *Betts* v. *Brady* continues to be used for proceedings like parole and probation revocation. (I am summarizing here the minimum constitutional requirements; actual practices in some jurisdictions may be more generous.)

The Sixth Amendment right to counsel extends before trial to preliminary hearings and other courtroom proceedings but not to traditionally secret grand jury proceedings, which remain securely under the control of the prosecutor. The right to counsel begins or "attaches" when the defendant first appears in court to answer a formal charge such as an indictment. Thereafter the police and prosecutors are supposed to communicate with the accused only through counsel and may not (for example) send spies to talk to him in the hopes of obtaining incriminating statements. Lawyers call this rule the "*Massiah* doctrine," after the name of the case in which it was first announced (*Massiah* v. *United States,* 1964).

If a criminal suspect is arrested before indictment, the Sixth Amendment does not apply and police questioning is permitted. The Fifth Amendment's privilege against self-incrimination does apply at this state of the proceedings, however, and the suspect may refuse to answer questions. Somewhat confusingly, the Supreme Court has held that a limited right to counsel can be derived from the Fifth Amendment. This is the source of the famous *Miranda* warnings: arrested suspects must be told that they can refuse to answer questions, that answers may be used against them, that they have a right to the assistance of a lawyer at any questioning and that a lawyer will be provided if they cannot afford one. In practice, a lawyer is not provided. By invoking the Fifth Amendment right to counsel, the suspect effectively terminates the questioning.

The problems of financing an effective right to counsel remain. Public defenders often have caseloads that are too heavy to allow effective preparation, and appointed private attorneys are usually compensated at rates far lower than those charged to paying clients. All defendants get lawyers unless they choose to represent themselves, but they do not necessarily get good lawyers. On rare occasions, convictions are overturned because a defense lawyer failed to provide "effective assistance," but the whole system would become completely unworkable if the courts seriously tried to insist on providing first-rate representation for the floods of drug sellers, robbers and rapists they are sending to prison.

A measure of equality is being achieved by leveling *down*. Federal laws against narcotics dealing and organized crime allow prosecutors to seize property obtained through unlawful activities and to freeze a defendant's assets before trial so that they will remain available for posttrial forfeiture. That means that a previously wealthy accused drug dealer or swindler may become instantly cash-poor and have to rely on a public defender. In 1989 the Supreme Court held that this practice does not deprive defendants of their constitutional right to counsel. Drug dealers whose assets are frozen are no more disadvantaged than any other class of indigents, reasoned the majority opinion of Justice Byron White. Those who cannot pay a private lawyer from honestly obtained funds must be satisfied with the services of the public defender.

The dissenting opinion by Justice Harry Blackmun pointed out a countervailing consideration:

> The "virtual socialization of criminal defense work in this country" that would be the result of a widespread abandonment of the right to retain chosen counsel, too readily would standardize the provision of criminal-defense services and diminish defense counsel's independence. There is a place in our system of criminal justice for the maverick and the risk-taker, for approaches that might not fit into the structured environment of a public defender's office, or that might displease a judge whose preference for nonconfrontational styles of advocacy might influence the judge's appointment decisions. (*Caplin and Drysdale* v. *United States,* 1989)

The equality principle implies uniformity of treatment and is furthered by impoverishing drug-addict robbers and millionaire drug importers alike. A free market approach makes room for the maverick and the risk-taker, but at the cost of making it inevitable that the kind of justice one gets depends on the amount of money one has. The same conflict of ideals is present in discussions of socialized medicine, or election campaign finance reforms, or proposals to provide "choice" through educational vouchers as an alternative to the public school monopoly. Past a certain point, the pursuit of equality tends to stifle individual initiative.

Gideon's Trumpet deserves its status as a legal classic, because it is a readable and well-informed account of a celebrated case. The book has also been widely used in schools and colleges to give students a picture of the American legal system. Therein lies a problem, because naive readers are likely to take the story not merely as an account of a single case but as a picture of the criminal justice system and the Supreme Court. A story can be true either as *description* (things happened this way on this occasion) or as *myth* (things happen like this in general). Lewis's account of Gideon's case is accurate as journalistic description, but at the mythological level it is highly misleading.

Take the protagonist. Gideon is a classic modern antihero who would be at home in a Woody Allen movie. He is a small-time gambler and ne'er-do-well who hurts mainly himself and his family. He is wrongly convicted of a minor property crime because he is too poor to hire a lawyer and is sentenced to a prison term reminiscent of that imposed on Jean Valjean. The implied moral? Convicts are harmless folks who are in prison because of some injustice and who can be freed without any great danger to society.

Take the lawyers. Abe Fortas is a prince of the Church of Law, designated by the pope to see that a peasant receives justice. Assisted by learned clerks, he speaks so eloquently that even the heathen are convinced. The case is then turned over to a lesser cleric named Turner, who wins the case with a brilliant cross-examination based on several days of dedicated preparation.

The supporting cast of lawyers also looks pretty good. The hapless counsel for the state is ineffective but dedicated and selfless. Prosecutors from other states join the convict's plea for justice. After the Supreme Court decides the case, bar association leaders dedicate their valuable time to making the constitutional right to counsel a living reality. The implied moral? Lawyers are virtuous souls whose efforts produce truth and justice.

Finally, there is the story of how the Supreme Court reinterprets the Constitution to bring it up to date. First, there was the revolutionary insight *(Powell* v. *Alabama)*, followed by the period of stagnation *(Betts* v. *Brady)*. The *Betts* doctrine gradually succumbed to its own inherent defects. Eventually practically everyone in the mainstream was ready to be convinced that the right to counsel should be absolute. When the time was ripe the Court unanimously announced the new rule, and genuine reform followed. The implied moral? The Supreme Court is a rational deliberative body, which can be trusted to lead the nation into a better future.

The mythological picture is not altogether false. There are relatively harmless defendants serving long sentences. Nowadays this is particularly likely to occur in small-time drug sales, where mandatory sentencing laws prevent judges from using common sense. There are lawyers at all levels of the profession who dedicate time to worthy causes. Sometimes they actually achieve truth and justice, or at least one or the other. The Supreme Court occasionally does perceive a new moral consensus and act upon it unanimously. The original school desegregation decision *(Brown* v. *Board of Education)* was roughly in the mold of *Gideon.* Defendants are sometimes innocent, lawyers are sometimes able and dedicated, and judges sometimes agree on good principles. But innocence, virtue and reasoned agreement are not exactly the norm.

Consider, for example, the Supreme Court's performance in the area of capital punishment. When the Court finally confronted the constitutionality of the death penalty in 1972, the justices wrote nine separate opinions. Although there was no majority opinion and no rationale, five justices voted to hold all existing death-penalty statutes unconstitutional. What kind of death-penalty statute the Court might have been willing to uphold was anybody's guess, but the central group of justices all said that death sentencing was unacceptably inconsistent in practice.

So any new death statutes had to constrain the judge and jury discretion that had led to inconsistency. Several states concluded that the way to do this was to set statutory conditions under which the sentencing authority "must" impose the death penalty. They got a rude shock when the Court held these mandatory statutes unconstitutional in the second round of death cases in 1976. A mandatory death penalty would not produce consistency, said some of the justices, because prosecutors, judges and juries would exercise discretion anyway. The correct way to control discretion was to bring it into the open and try to guide it with a laundry list of factors the sentencing authority ought to take into account.

In subsequent cases the justices have struggled to apply various doctrines that are inconsistent in principle. On the one hand, jury discretion must be controlled to prevent inconsistency and discrimination. On the other hand, the jury must be free to grant mercy on any basis it sees fit. On the one hand, only murderers and not (say) sadistic rapists deserve death. On the other hand, doctrines like the felony murder rule and conspiracy may be used to extend capital punishment to persons who did not personally kill or intend to kill.

Whatever one thinks of the death penalty, the Supreme Court's handling of the issue has been irrational. The problem is not that the justices are individually incapable of reasoning logically but that they are in the business of deciding questions (abortion, affirmative action, medical ethics) about which well-informed persons passionately disagree. It is unsurprising that the Court has become as divided on these subjects as the nation.

It is also easy to tell a story of crime and lawyers that reflects the bizarre side of the American legal culture of the late twentieth century. Take the saga of Robert Alton Williams, the antihero of a judicial puzzle piece known to legal scholars as the "Christian Burial Speech" case. The story begins on Christmas Eve of 1968, when a ten-year-old girl named Pamela attended an event with her family at the Des Moines, Iowa, YMCA. She excused herself to go to the bathroom and never came back. Shortly afterwards Williams was seen hastily leaving the building with what appeared to be a body wrapped in a blanket. He drove east across Iowa to Davenport, discarding Pamela's clothing on the way and hiding the body somewhere in the countryside.

On December 26 Williams phoned a Des Moines lawyer named McKnight and then surrendered to the police in Davenport on McKnight's advice. McKnight went to the Des Moines police station to arrange matters, and there he had another telephone conversation with Williams in the presence of police officers. He told his client that a Captain Leaming would be coming to Davenport to take him back to Des Moines by car, that Leaming would not question him during the trip and that he should not discuss the case with anyone until he could confer with McKnight. McKnight also told Williams he would have to reveal the location of Pamela's body after his return to Des Moines. Although no one testified that Captain Leaming or other officers made any explicit promise not to question Williams, various judges later inferred that the police had promised, perhaps by their silence, to abide by McKnight's terms.

Leaming and another officer went to Davenport and picked up Williams, who had been arraigned in court there on a murder warrant and repeatedly advised of his constitutional rights. Shortly after leaving Davenport on the 160-mile return trip, Leaming delivered the notorious "Christian burial speech." He addressed Williams (an escaped mental patient with strong religious tendencies) as "Reverend" and urged him to ponder the desirability of stopping en route to locate the body (before anticipated snowfall could conceal it) so the girl's

family could give her "a good Christian burial." Although the speech
was not in the form of a question or demand for information, it clearly
was intended to appeal to Williams's conscience and influence him to
show the police where he had hidden the body. It had the intended
effect. Some hours later, as the car approached the Des Moines area,
Williams took the officers to the body hidden in a ditch about two
miles off the interstate.

State courts affirmed the resulting conviction for murder, but a
federal appeals court granted a writ of habeas corpus and ordered a
new trial. The defendant's *Miranda* rights were violated, said the
opinion, when Leaming in effect questioned him after he and his
attorney had invoked his right to counsel and privilege against self-
incrimination. The evidence that he had led the officers to the concealed body should therefore not have been admitted.

When the state of Iowa took the case to the Supreme Court in 1976,
law enforcement advocates hoped that the Court might seize the
occasion to overrule the controversial *Miranda* decision itself. *Miranda* had been a 5-4 decision in 1966, and Republican presidents had
appointed several new justices in the meantime. What these hopes
overlooked was that Williams had retained counsel and had appeared
in court before the crucial car trip. The Supreme Court held that
therefore judicial proceedings had already begun, and so the Sixth
Amendment right announced in the *Massiah* decision was applicable.
The 5-4 majority opinion on this basis bypassed the *Miranda* issue,
reversed the conviction on the authority of *Massiah* and sent the case
back for a new trial in the state courts (*Brewer* v. *Williams,* 1977).

Just as Gideon had threatened to turn his great case into farce by
representing himself at the retrial, Williams's new lawyers adopted a
defense theory that took no advantage of the Supreme Court decision.
Having obtained forensic reports that were previously unavailable,
they noticed that a component of sperm had been found on Pamela's
body, but there was no indication that spermatozoa were present. The
rapist-murderer might therefore have been sterile, but Williams was

not sterile. Perhaps Pamela was killed by another person, who had left the body in Williams's room at the YMCA. Williams then panicked, rightly assuming he would be blamed, and fled with the body in a futile attempt to save himself.

The adversary/accusatorial system of criminal justice permits a defense lawyer to float a suggestion like that by innuendo and imagination alone, without putting the defendant on the stand to deny that he committed the murder. In this trial the work was so skillfully done that the trial judge anticipated a possible acquittal and made plans to whisk Williams out of town quickly in case a lynch mob formed. The jury again convicted, however, and the state supreme court again affirmed. Appointed attorneys took the case back to the federal courts, and the federal court of appeals overturned the conviction a second time.

This time the problem was the victim's body. The jury at the retrial had not been told that Williams led police to the body, but the prosecution was allowed to use the body to establish the *corpus delicti* of the crime—that is, that the victim died by violence. Of course the police found the body only after Williams had led them to it, and so it was itself the product of a chain of events set in motion by the unconstitutional Christian burial speech. The state argued that search parties would eventually have found the body anyway, but the federal court said that did not matter. Captain Leaming had acted in bad faith by trying to get around Williams's constitutional rights with an indirect interrogation, and any evidence thereby obtained was irretrievably tainted.

Imagine what a third prosecution (in 1983 or later) would have looked like if that ruling had stood. The state might have tried to establish the *corpus delicti* with testimony that Williams left the YMCA shortly after the crime with what looked like a body. The defense could have responded that since for legal purposes Pamela's body was never found, the court must assume that she might be well and alive somewhere at the age of twenty-five. The citizens of Iowa were spared this travesty when the Supreme Court granted a petition for review a second time and reinstated the conviction. A Sixth Amendment violation requires only exclusion of evidence that would not have been discovered in any case, reasoned the majority, and the state courts had found that the searchers would eventually have found the body (*Nix* v. *Williams,* 1984).

What moral lessons could we draw from the long saga of the Christian Burial Speech case? First, many criminal defendants are very dangerous people who have committed unspeakable crimes. Second, the purpose of giving defendants lawyers is not necessarily to bring out the truth. Suppressing evidence and confusing juries is not unethical practice by our standards but good defense lawyering.

Whether lawyers achieve justice depends on what we mean by that term. In liberal thinking, the concept of "justice" tends to collapse into the category of "rights." The Constitution is the supreme law of the land, and so it takes precedence over local regulations such as the prohibition of murder. The most important part of the Constitution is not the limitation of federal government power or the grant of legislative power to the Congress but the Bill of Rights. It follows that courts in criminal cases should protect the rights of the defendant and subordinate all other considerations to that end.

From this standpoint it is not an "injustice" if a man escapes punishment after raping and murdering a child. The victim is dead, after all, and nothing can be done for her. Her friends and family have no "rights" in a criminal prosecution, which is a contest between the state and the accused. If the legal rights of the parties were observed, justice was done. Those who think justice demands swift and certain punishment for murder are regarded as vindictive, and also as irrational for refusing to acknowledge that the real causes of crime are social injustice and mental illness.

Rights advocates generally assume that something *will* be done, one way or another, to protect the public from truly dangerous individuals. Justice Thurgood Marshall, for example, hinted that the state should commit Williams to a secure mental institution, presumably on the same evidence of murder he wanted to exclude from the criminal prosecution. But the insane also have their rights advocates, who have succeeded in emptying the mental institutions. Today persons who would formerly have been institutionalized fill the streets and parks of our cities.

I have come to praise *Gideon's Trumpet,* not to bury it. It is a legal classic because it captures the events and ideas of a particular period in history. Although its limits must be kept in mind, it remains a remarkable achievement of legal journalism. I read it first nearly thirty years ago, and read it again with pleasure just now. But one thought kept recurring in my mind.

1964 sure was a long time ago.

15

Left Behind

FOR A RESIDENT OF THAT FAMOUSLY LEFTIST UNIVERSITY CITY CALLED Berkeley, California, one of the most remarkable features of the 1990s is the virtual disappearance of the Left as a unified political force. In The Twilight of Common Dreams: Why America Is Wracked by Culture Wars *(Metropolitan Books/Henry Holt, 1996) Todd Gitlin, one of the most prominent of Berkeley's leftist academics, describes and bemoans the twilight of his own dreams. My essay was first published in* Books & Culture, *September/October 1996.*

Todd Gitlin begins his book on our current culture war by recounting the absurd Oakland (California) school textbook battle of 1992. He presents this spectacle as a paradigmatic example of the retreat of the Left from universalism and its descent into tribalist irrationalism.

The state board of education had approved for local adoption a new series of social science textbooks for kindergarten through eighth grade. The books were written and edited for Houghton Mifflin by a group of consultants headed by UCLA history professor Gary Nash, a leftist of the same stripe as Todd Gitlin. The series was on the whole proudly multiculturalist, placing great emphasis on the positive contributions of indigenous people and minority races and the crimes that were committed against them.

No one would have been surprised if the textbooks had been denounced by Rush Limbaugh and Lynne Cheney, but on this occasion the cultural right was silent. When the minority-dominated Oakland school board considered purchasing the books, however, the public hearing was packed with racial demagogues of the Left, who searched out isolated passages that could be interpreted as offensive. The most aggressive attacker was a professor of ethnic studies from San Francisco State University. She appeared with a platoon of her students, who probably received academic credit for this laboratory experiment in agitprop. Gitlin quotes another long-time white leftist as remarking that the attackers "would have spoken in the same vein if the authors [of the textbooks] had been George Wallace, Ross Barnett, and Bull Connor." Why not? George Wallace and Gary Nash are both white.

In the end the series was rejected, and Oakland teachers went into the next school year without any social science textbooks at all. What was tragic about this orgy of racialist symbol-mongering, according to Gitlin, was that it occurred while the public schools were grossly underfinanced and amid "a stupefying degree of inequality in American society and, in particular, among African Americans." Instead of organizing against "rock-bottom inequalities and racial discrimination," however, the activists of identity politics chose to fight "real and imagined symbols of insult."

Gitlin, a Berkeley sociologist who recently moved to New York University, was a founding member of what used to be called the New Left. This was the energetic student movement of the late 1960s that opposed the Vietnam War and racism while glorifying a personal hedonism based on "sex and drugs and rock-and-roll." Critics of the

Sixties Left (it can hardly be called "new" today) see it as a self-indulgent and self-righteous Children's Crusade that began with some correct insights—Martin Luther King Jr. was right on civil rights and Lyndon Johnson was wrong on Vietnam—but then went haywire about everything else. The main theoretical contribution of the Sixties Left was "participatory democracy," a recipe for anarchy that destroyed every organization that put it into practice. Its continuing legacy is the suffocating piety known as "political correctness."

Unrepentant veterans like Gitlin concede that something went terribly wrong, but they maintain that the Sixties Left was founded on a noble vision of universal equality which they would love to recapture. That seems impossible, however, because critics and nostalgic veterans alike agree that, for better or worse, the Left today is hopelessly fragmented. Gitlin goes so far as to say that "today it is the Right that speaks a language of commonalities. Its rhetoric of global markets and universal freedoms has something of the old universalist ring. To be on the Left, meanwhile, is to doubt that one can speak of humanity at all."

In place of a unified humanity, today's Left celebrates what sociologists call an "identity politics" of discrete groups pursuing their own goals. The principal groups are racial minorities, feminists and gays, with some others like environmentalists and disabled people as associate members. The chant on the campuses excoriates not capitalism but "racism, sexism and homophobia." The class enemy is no longer the millionaire or the militarist but the infamous straight white male, including even straight white male sociology professors who plead in extenuation their long record of marching in support of Left causes. Understandably, Todd Gitlin is not comfortable with a Left whose idea of fighting racism and sexism is to typecast people like himself as oppressors.

If, as Gitlin charges, the Left has abandoned the universalist ideal of socialism and substituted a self-defeating tribalism that cedes the moral high ground to the market worshipers, what is the remedy?

Gitlin knows very well that the activists of identity politics will not change their course in response to the pleadings, however eloquent, of a straight white male academic who feels left out. Rebuilding after a political catastrophe requires actions, probably drastic ones. Gitlin consistently refuses to follow through on the obvious implications of his own analysis, however, probably because by doing so he would instantly reclassify himself as a neoconservative. I will point out just two examples.

First, should changes be made in government policies that actively encourage identity politics? Gitlin reports that the census employs a system of racial categorization "that is rigid to a degree that astounds (and horrifies) many people outside the United States, especially in countries like Canada, Mexico, and France, which have banned the collecting of racial statistics." New racial categories are added from time to time "as a result of political pressure from groups seeking to maximize their representation in public life." The categories affirm the basis for identity politics and provide "a tactic for garnering resources." Remember that professor of ethnic studies and her student shock troops.

In short, identity politics flourishes in part because government policies encourage and subsidize it. Gitlin's narrative clearly implies that those foreigners are right to be horrified. Should we, then, do as the foreigners do and stop classifying people on the basis of rigid racial categories? Gitlin must be familiar with current proposals to allow people to classify themselves on the census as "multiracial," but he avoids the subject. He does not say whether the legal system should attempt to return to Martin Luther King's principle—now quoted mainly by conservatives—that people should be judged not by their race but by the content of their character. Liberals used to defend distributing benefits by race as a temporary measure, to be abandoned when racism has been sufficiently reduced, but that is senseless if racial entitlements are worsening the very conditions they are sup-posed to be curing. I surmise that while he cannot endorse moving toward a color-blind legal system and still call himself a leftist, Gitlin would not be displeased if the Right were to take that step and incur the blame.

Second, should disillusioned leftists reconsider their hostility to traditional religion and even see religion as a promising basis for a universalist politics of human equality? Gitlin has a nostalgic chapter on Marxism, which he correctly describes as "a theology without God." He refers to theistic religion only dismissively, however, implying the usual Left prejudice that a serious interest in God is inherently reactionary. Gitlin also dismisses out of hand the theory that one important cause of social pathology is the breakdown of the traditional family, a circumstance that is linked to the abandonment of religious morality.

The allure of Marxism was that it bridged the fact-value distinction, painting a total picture of reality in which a passionate social utopianism seemed to be backed by a hardheaded materialist science. Marxism is now moribund not only because of its spectacular failures in practice but also because its theory does not address the issues that preoccupy the contemporary Left. Who goes to Marx for guidance on gay marriage or the evils of Eurocentrism? The doctrine that economics is the base and culture the superstructure, like the doctrine that justice should be color-blind, is more often found on the *Wall Street Journal* editorial page than in radical journals. Even leftist economists appreciate the utility of free markets for creating wealth, and Gitlin concedes that the international nature of the economy makes it impossible for a national government to redistribute wealth drastically without provoking a disastrous flight of capital. The only presidential candidate in 1996 who did not support market internationalism was Pat Buchanan.

What the Left plainly needs is a new theology, with or without God. Gitlin makes clear what the elements of such a theology must be. It must provide a universal vision that inspires people to regard themselves as fundamentally united, despite their differing social circumstances and cultural experiences. It must provide a basis for an objective rationality of both fact and value, refuting the current Left doctrine that "objectivity is only another word for white male subjectivity." It must reject the market-oriented notion that individual gratification is the purpose of life, by providing a higher purpose. It must provide a reason for the economic winners to be generous and compassionate and for the losers to strive to become as productive as they are able.

Where is such a theology to be found? I could offer a suggestion, but I don't think Todd Gitlin wants to hear it.

16

Pomo Science

---

*THIS ESSAY CONTRASTS THE RATIONALIST AND RELATIVIST APPROACHES
to science—and to biblical studies—in its review of* The Science Wars, *a
special issue of* Social Text *(14, nos. 1-2 [1996]), and John Horgan's*
The End of Science: Facing the Limits of Knowledge in the Twilight of
the Scientific Age *(Helix Books/Addison-Wesley, 1996). It was originally
published in* Books & Culture, *November/December 1996.*

New York University physicist named Alan Sokal played a
cruel practical joke in 1996 on the editors of the postmodernist journal *Social Text*. "Pomos," as postmodernists are
not-so-affectionately called by other academics, are noted for leftism
in politics, relativism in epistemology and pretentious murkiness in
expression. Pomo writing is radically skeptical about the objectivity
of knowledge, including scientific knowledge. This has led mainstream scientists to denounce the Pomos as enemies of science, far
more dangerous than the despised creationists because they hold
influential positions in universities.

Alan Sokal is himself a leftist, proud of his stint teaching under the Sandinistas in Nicaragua, but he is a rationalist—much like the sociologist Todd Gitlin (see the preceding essay). To demonstrate that Pomos are pretentious phonies who give the Left a bad name among sensible people, Sokal stitched together an incoherent article that combined quotations from Pomo authors (including some of the editors of *Social Text*) with nonsensical scientific analogies. Then he ponderously titled it "Transgressing the Boundaries: Towards a Transformative Hermeneutics of Quantum Gravity," signed his name and title, and sent the monstrosity off. The editors, pleased to be taken seriously by a real scientist, published the article in a special issue titled *The Science Wars,* which had been meant to rebut their rationalist critics. Sokal then turned the Pomo counterattack into a debacle by gleefully revealing his hoax to the press in the May/June 1996 issue of the journal *Lingua Franca.*

The fun begins with Sokal's preposterous title and continues throughout his brilliant parody, but in the endnotes he really lets rip. You can almost hear him cackling as he crafted these notes, with their superb mimicry of Pomo pieties, their interweaving of genuine references and fabricated sources, and their inside jokes (many of which will doubtless be accessible to only a handful of readers), all conducing to a delicious absurdity. (The very first item in the reference list is the infamous piece by Hunter Haveline Adams III from *African-American Baseline Essays,* which Sokal cites with a straight face.) Here is a sample of Sokal's endnote style:

54. Just as liberal feminists are frequently content with a minimum agenda of legal and social equality for women and "pro-choice," so liberal (and even some socialist) mathematicians are often content to work with the hegemonic Zernelo-Fraenkel framework (which, reflecting its nineteenth-century liberal origins, already incorporates the axiom of equality) supplemented by the axiom of choice. But this framework is grossly insufficient for a liberatory mathematics, as was proven long ago by Cohen, 1966.

Not since Vladimir Nabokov's *Pale Fire* have the notes to a text so richly rewarded close attention.*

Commentators from the left and right of the spectrum jumped at the opportunity to ridicule the embarrassed Pomos. The editors made themselves look still worse by their response, saying among other foolish things that the article's "status as a parody does not alter substantially our interest in the piece itself as a symptomatic document." That confirmed Sokal's point that a parody of Pomo science-talk is hard to distinguish from the real thing. The notoriously nihilistic Stanley Fish—who is executive director of Duke University Press, which publishes *Social Text*—added to the merriment by piously warning that Sokal's resort to deception would damage the relationship of trust that presently prevails in academic affairs.

Sokal's prank gave a black eye to an interdisciplinary movement called "science studies." Science studies is something like anthropology, focusing on the cultural and political aspects of the scientific profession. It includes everything from mainstream social scientists to literary theorists looking for texts to deconstruct. Some resentful scientists dismiss the whole field as a pack of English majors who couldn't pass freshman calculus but who presume to treat the scientific community as if it were a primitive tribe with a colorful mythology. Evelyn Fox Keller, one of the more respectable feminist theorists in science studies, complains that her scientific colleagues "readily confuse the analysis of social influence on science with radical subjectivism, mistaking challenges to the autonomy of science with the 'dogma' that there exists no external world."

More discriminating scientists concede that the culture of the scientific community is a legitimate subject of study and that scientific priorities are sometimes skewed by social factors like money and politics. But they also insist that science continually tests its theories

---

*The full text of Sokal's parody is available on his web page: http://www.nyu.edu/gsas/dept/physics/faculty/sokal/index.html

against external reality by experiment, so that, unlike literary studies or philosophy, science produces a continually growing body of reliable, transcultural knowledge. As the physicist Steven Weinberg expressed it (in a fine essay on the Sokal hoax in *The New York Review of Books* [August 8, 1996, pp. 11-15]), "If we ever discover intelligent creatures on some distant planet and translate their scientific works, we will find that we and they have discovered the same laws." Scientists may have their prejudices, but these are relatively unimportant if they do not prevent science from progressing steadily toward a truth that is the same for everybody.

Biblical studies provides an analogy. The writers of the four Gospels have differing viewpoints and interests, and this is an interesting subject for scholars. But for believers such matters are insignificant compared with the fact to which the Gospels all testify, the resurrection of Jesus. If the resurrection is a *fact,* then what that fact implies is far more important than the cultural setting of the reporters. When a naturalistic age reinterprets the resurrection as a myth, however, the mythmakers and their audiences become the main story. For the critics Jesus himself—the so-called historical Jesus—then becomes something like King Lear. A man by that name may have existed, and perhaps he even had trouble with his daughters, but the Lear we know is the creation of Shakespeare and his culture.

John Horgan's fascinating collection of interviews with leading scientists provides a very different model of how literary intellectuals might write about scientists. Horgan was an English major who abandoned literary criticism as pointless because it generates nothing but an endless variety of conflicting interpretations. He gravitated toward science as an activity that addresses questions that actually can be answered and became a highly regarded writer for *Scientific American.* Far from challenging the objectivity of science, Horgan thinks its very success in discovering universal truth jeopardizes its future. The time is coming, he says, when the big questions that *can* be answered will have been answered. What will remain is details—

filling in the pieces—and speculative theories invoking mathematical entities like superstrings whose physical existence may never be empirically testable.

It is best not to take too literally this "end of science" thesis, which Horgan used as a conversation opener to give his interviews a common focus. Horgan understands that science still has major puzzles to solve, including the origin of life, the nature of consciousness and the composition of cold dark matter. His claim is primarily that those puzzles will be solved within the boundaries of present theories without the need for revolutionary new discoveries. In curiously religious language, Horgan explains,

> My guess is that this narrative [the standard scientific materialist understanding of cosmic and biological evolution] that scientists have woven from their knowledge, this modern myth of creation, will be as viable 100 or even 1000 years from now as it is today. Why? Because it is true. Moreover, given how far science has already come, and given the physical, social, and cognitive limits constraining further research, science is unlikely to make any significant additions to the knowledge it has already generated. There will be no great revelations in the future comparable to those bestowed upon us by Darwin or Einstein or Watson and Crick.

Again, Christian theology provides a rough analogy. With the incarnation, God has spoken definitively. The Holy Spirit still has a lot to do, but Christians expect no comparable future revelation, and we certainly do not expect revealed truth to be replaced by something substantially different. Science and theology thus agree that there is a fundamental reality behind the changing patterns of language and culture and that true knowledge of that reality endures even though various interpretations are culture-bound. Horgan's prime example of a permanent truth is the Darwinian theory of evolution; I predict that Jesus Christ will be a living reality long after Darwinism has been relegated to the history curriculum.

There is a lot of middle ground between the "end of science" thesis and the "science is just another tribal belief system" thesis. Of course science can transcend cultural differences to generate objective knowledge on some subjects. Rationalists like to point out that not even Pomos want to fly in an airplane designed by a committee picked for its multicultural diversity. Such examples can be misleading, though, if applied too broadly. Science is determined to explain *all* aspects of reality, and that ambition sometimes tempts scientists to theorize extravagantly from part of the evidence while ignoring or explaining away the facts that don't fit the theory. When scientists do that, they really are culture-bound producers of texts.

Our children can look forward to finding out whether the major components of the modern myth of creation are as permanent as Horgan thinks, or whether the twenty-first century will experience not the end of science but the transformation of science.

17

Harter's Precept

I AM CONVINCED THAT CONSCIOUS DISHONESTY IS MUCH LESS IMPOR-
tant in intellectual matters than self-deception. The great physicist
Richard Feinberg loved to warn beginning scientists, "The first prin-
ciple is that you must not fool yourself, and you are the easiest person
to fool." (See my Defeating Darwinism by Opening Minds *[InterVar-*
sity Press, 1987], p. 46.) Bruno Müller-Hill teaches the same maxim:
"First you fool yourself, then you fool others." In The Social Misconstruction of Reality: Validity and Verification in the Scholarly Community *(Yale University Press, 1996), Richard F. Hamilton shows his*
readers that some prominent social scientists have done just that. My
essay was first published in Books & Culture, *March/April 1997.*

The German biologist Bruno Müller-Hill tells a memorable story to illustrate his thesis that "self-deception plays an astonishing role in science in spite of all the scientists' worship of truth":

> When I was a student in a German gymnasium and thirteen years old,
> I learned a lesson that I have not forgotten. . . . One early morning our

physics teacher placed a telescope in the school yard to show us a certain planet and its moons. So we stood in a long line, about forty of us. I was standing at the end of the line, since I was one of the smallest students. The teacher asked the first student whether he could see the planet. No, he had difficulties, because he was nearsighted. The teacher showed him how to adjust the focus, and that student could finally see the planet, and the moons. Others had no difficulty; they saw them right away. The students saw, after a while, what they were supposed to see. Then the student standing just before me—his name was Harter—announced that he could not see anything. "You idiot," shouted the teacher, "you have to adjust the lenses." The student did that and said after a while, "I do not see anything, it is all black." The teacher then looked through the telescope himself. After some seconds he looked up with a strange expression on his face. And then my comrades and I also saw that the telescope was nonfunctioning; it was closed by a cover over the lens. Indeed, no one could see anything through it. ("Science, Truth and Other Values," *Quarterly Review of Biology* 68, no. 3 [September 1993]: 399-407)

Müller-Hill reports that one of the docile students became a professor of philosophy and director of a German TV station. "This might be expected," he wickedly comments. But another became a professor of physics, and a third a professor of botany. The honest Harter had to leave school and go to work in a factory. If in later life he was ever tempted to question any of the pronouncements of his more illustrious classmates, I am sure he was firmly told not to meddle in matters beyond his understanding.

One might derive from this story a satirical "Harter's Precept" to put alongside Parkinson's Law (bureaucracy expands to the limit of the available resources) and the Peter Principle (one rises in a hierarchy up to one's level of incompetence). Harter's Precept says that the way to advance in academic life is to learn to see what you are supposed to see, whether it is there or not. As Sam Rayburn used to explain to new members of Congress, you've got to go along to get along.

Richard Hamilton's *The Social Misconstruction of Reality* indicates that many social scientists seem to have guided their careers by the light of Harter's Precept. Hamilton gives three major examples of erroneous theses that gained the status of fact in social science despite the absence of evidentiary support: (1) Max Weber's thesis that the Protestant Ethic spurred the advance of capitalism, (2) the widely accepted thesis that Hitler's main electoral support came from the lower middle classes (the despised petit bourgeoisie of Marxism) and (3) Michel Foucault's thesis that the modern prison evolved not as a more humane alternative to the cruel physical punishments of earlier centuries but as part of a wide-ranging scheme by sinister forces to enforce a pervasive social conformity.

Persons who want to follow the fact-heavy specifics of the three cases should read Hamilton. Very briefly, Weber relied on slender evidence and based a major point on a typographical error. He thus offered an elaborate explanation for a phenomenon—the supposed domination of the business economy by Protestants—that the data did not support. The claim that Hitler obtained his electoral support disproportionately from the lower middle class arose from ideology rather than evidence and became generally accepted because influential scholars uncritically repeated it. Actual analysis of voting records published in the 1980s showed that the lower middle class did not provide exceptional support for Hitler, his support being greatest among the more affluent groups.

The easiest of Hamilton's three case studies to summarize, and the one with the greatest current impact, is that of Michel Foucault. Foucault's book *Discipline and Punish* argued that the modern prison replaced the barbaric punishments of earlier ages because a malevolent "they" decided that brutality is less effective than a more superficially humane punishment "that acts in depth on the heart, the thoughts, the will, the inclinations." To accomplish this pervasive control, "they" adopted Jeremy Bentham's "panopticon" design for the prison, which permits guards to watch every solitary prisoner in every cell at all times. Such controls were then extended "throughout the social body" to create a "disciplinary society" of domination.

The factual basis of Foucault's thesis consisted mainly of offhand assertions and grotesque errors, supported with references to obscure sources that didn't really support the text. For example, Bentham's panopticon was an eccentricity that was rarely copied and never successful; most modern prisons are built on the entirely different Auburn plan. Nineteenth-century governments preferred to transport felons to distant penal colonies rather than to keep them constantly under surveillance. Foucault's method was a freewheeling interpretation of selected texts, and he made little effort to situate the texts in context or to distinguish eccentric proposals and exceptional events from regular practice.

These methodological blunders escaped the notice of the eminent reviewers who praised Foucault's history and helped make him fashionable in the academy. The reviewers also failed to notice the transparently paranoid nature of Foucault's attribution of imaginary evils to unspecified malevolent forces. Just as Weber's thesis was revered by sociologists but ignored by historians who knew the facts, Foucault is a cult hero in the humanities but not taken seriously by penologists.

Yet the cases of Weber and Foucault are otherwise very different. Müller-Hill warns us that even scientists who worship truth are prone to deceive themselves, but at least they are embarrassed when their errors come to light—as Weber would have been. Disciples of Foucault, on the other hand, are typically indifferent to criticisms of their master's accuracy. This is not surprising, because Foucault himself was a nihilist obsessed with power, who believed that "truth" is something imposed by powerholders, not something found by impartial investigation.

Foucault was also personally a sociopath who sought the limits of experience in drug use and in the sadomasochistic depths of San Francisco's gay bathhouses. (He died of AIDS in 1984.) His use of deliberate or reckless errors in his scholarship may have been another attempt to test the limits, just to see what he could get away with.

Foucault's writing is not about objective reality, which to him was a form of tyranny, but about his own obsessions and his determination to defy every restraint. That such an enemy of reason became one of the most influential and honored of late-twentieth-century philosophers dramatically refutes his own thesis that we live in a "disciplinary society." (Why did "they" tolerate Foucault?) It tends rather to suggest that parts of our universities are going through what Plato described as the last stages of democracy, when all restraints break down and nihilism is both preached and practiced.

Hamilton concludes with an analysis of how major scholarly errors get made and perpetuated: theorists fall in love with their theories, the appearance of scholarship can often pass for the reality, and clever charlatans like Foucault can evade the scrutiny of specialists and appeal directly to those who are all too willing to deceive themselves. Above all, there is the wisdom of Harter's Precept. Revealing that the lens cap is still on the telescope isn't necessarily good for one's career.

18

The Circus of Death

SCIENTISTS RIGHTLY TAKE PRIDE IN THE FACT THAT SCIENCE IS A SELF-correcting enterprise, in which errors on important matters are bound to be exposed. Most of the time the critical apparatus works, but under extraordinary circumstances it may fail. Political and financial pressures may affect the objectivity of leading researchers, and crucial concepts may be defined so vaguely that they elude rigorous testing. Some of us think this is what has happened in the case of AIDS. The following essay, a review of Elinor Burkett's The Gravest Show on Earth: America in the Age of AIDS *(Houghton Mifflin, 1995; the review originally appeared in the March 1996 issue of* First Things*), was written before the advent of the new "drug cocktails" (which still employ AZT) and the resulting claims of medical miracles. I'll leave it to the future to decide whether my skepticism was mistaken.*

Elinor Burkett portrays the AIDS culture wars as a kind of circus in which a series of clowns and villains perform on stage while the audience slowly dies from neglect. She does not purport to give the complete story of AIDS. Her book has nothing to say about

Africa or Asia and very little about hemophiliacs, intravenous drug addicts or infants. Burkett's story is about the relationships between gay activists, scientists, drug companies and the media—as seen by a Miami-based reporter who seems to have become disgusted with just about everybody.

The theme of the book as a whole is encapsulated in the first chapter, a dual biography of two flamboyant egotists who became allies: biomedical scientist Robert Gallo and gay playwright Larry Kramer, founder of ACT-UP. Gallo was determined to prove that a human retrovirus discovered in his laboratory at the National Cancer Institute was responsible for some disease serious enough to merit a Nobel Prize for the discoverer. Kramer played the role of spokesman for gay rage, freely accusing scientists and government officials of sponsoring a "holocaust" or "genocide" because they had not taken swift enough action to prevent or cure AIDS.

The accusations were absurd, but the liberal media culture found it easy to believe that the hated Reagan administration was uninterested in curing AIDS because it didn't care about the victims. In this climate of opinion, health officials were under enormous pressure to do something—*anything*—to show that they had a viable program. The immediate result was the famous press conference in April 1984 at which the secretary of health and human services announced on shaky evidence that one of Gallo's retroviruses (later named HIV) was the "probable" cause of AIDS. The secretary spectacularly rebutted the barrage of criticism with the prediction that with the pathogen known, a vaccine would follow in a couple of years. The resources of government were soon devoted to exploiting the breakthrough. AIDS research became HIV research, and funding for studying other possibilities became unavailable.

The second major step taken to stem the criticism was the approval of AZT as a treatment. Burroughs Wellcome, the manufacturer of AZT, took advantage of the panic to rush through Food and Drug Administration approval of a drug that it had on the shelf as a failed cancer cure. Activist demonstrators, with help from conservative politicians pushing an agenda of deregulation, intimidated FDA scientists who found the evidence of efficacy inadequate. The FDA, which had stoutly resisted the claims of terminal cancer patients in the laetrile episode, caved in to the far more potent AIDS lobby. Whether AZT gives any substantial benefit to AIDS sufferers is still hotly debated, but nobody disputes that it has immensely benefited its manufacturer.

Burkett never pins down exactly what she is angry about. No doubt AIDS politicking has involved lots of greed, paranoia and hypocrisy, but that doesn't distinguish AIDS from a lot of other things going on in government and business. The promised vaccine or cure is nowhere in sight, but the partnership between gay activists and scientists led the government to provide "a level of Federal attention, support and services that were the envy of people with cancer, multiple sclerosis, and heart disease." Progress has been made treating some AIDS-related conditions, and new antiviral drugs that might be better than AZT are on the way. That AIDS has proved so difficult to cure or even understand is cause for disappointment, but where's the scandal?

Burkett actually gives her readers the facts needed to answer that question, but it takes a bit of background to separate the wheat from the chaff. More than highlighting the faults of individuals, the AIDS failure illustrates the adage that it's no advantage to travel faster unless you are going in the right direction. In science as in law, meticulous attention to established procedure is essential for accuracy, but it is also extremely expensive, especially when delay may cost lives. Scientists, like other people, tend to see what they want to see, and they have no immunity to traits like impulsiveness or greed. Ordinarily these traits are held in check by rigorous peer review before claims even get published and by the still more rigorous criticism that really important claims will receive from other laboratories for months or years thereafter.

If government officials had waited for Robert Gallo's virus hypothesis to survive that kind of criticism before making it the basis of

a huge research program, however, they would have been excoriated
for going about business as usual while multitudes were dying. The
path of least resistance was to go for broke with the best idea currently
available, and that is why the campaign against "the virus that causes
AIDS" was announced at a press conference before the scientific
community had a chance to evaluate whether the virus really does
cause AIDS.

Once the scientific army was on the march, with everybody's
research program based on the premise of HIV causation, the time for
debating the starting point had passed. When the prestigious Berkeley
molecular biologist Peter Duesberg published a review article in 1987
opposing the HIV hypothesis, he was ignored for two years and then
ridiculed and deprived of all funding. If the same arguments had been
widely circulated in the spring of 1984, they would have received a
respectful hearing, and the course of events might have been entirely
different.

The same sense of urgency accounts for the hasty approval of AZT
and the drastic easing of previously established drug-testing procedures. Drug tests must be placebo controlled and "double-blind,"
meaning that neither the doctors nor the patients know who is getting
the drug and who the placebo. They also should be continued long
enough so that long-range as well as short-term effects are measured,
and repeated so it is clear that different research teams get similar
results. These precautions are important because of the notorious
"placebo effect" on patients and because doctors who understandably
want the drug to work may see benefits that aren't really there or are
only temporary.

Rigorous design and evaluation of trials is particularly necessary
because the drug companies that are seeking approval fund the studies
and even select the researchers. Scientists may have a lot to gain by
putting the most favorable possible spin on the data, especially when
(as in the case of AIDS) their superiors in the scientific hierarchy
desperately want to announce the discovery of an effective treatment.

Independent evaluators need to make sure that unfavorable interpretations of the data were not overlooked.

Burkett correctly reports that the crucial study that purported to establish the efficacy of AZT adhered to standard criteria only in a formal sense. Patients and doctors knew who was getting AZT, because the drug caused such severe anemia (among other unpleasant effects) that many patients required blood transfusions to survive. After seven months, however, nineteen members of the placebo group had died as opposed to only one in the AZT group. The trial was then halted before completion on the ground that efficacy had been established and it would be unethical to continue giving any AIDS patient a placebo. Three months later, thirteen of the placebo group who had been switched to AZT had died, along with seven who had been on AZT from the beginning. Despite serious reservations among FDA staff over what this ambiguous trial had proved, the FDA gave in to pressure and speedily approved AZT for the market.

When the later and far more extensive European Concorde trial showed that AZT did not extend life for asymptomatic HIV positives, enthusiasm was dampened but not extinguished. AZT remains in very widespread use for both AIDS patients and asymptomatic HIV positives. Because it is now considered unethical to include a placebo group in tests, AZT is also the "gold standard" against which other drugs are compared in efficacy tests. As a result AIDS patients now have the dubious benefit of being able to take a variety of drugs with uncertain benefits and damaging side effects.

Almost twelve years after the 1984 press conference, knowledgeable AIDS patients have learned to distrust press releases about impending wonder drugs. Vaccine trials have been put off into the indefinite future, amid confusion over exactly what a virus should be expected to do. In a statement reported in *The New York Times* on December 6, 1995, Robert Gallo remarked, "The No. 1 reason that we don't have a vaccine today is that we don't know what to induce in humans." Such statements reflect a significant gap in the theoretical knowledge needed to design a treatment strategy, and indeed scientists are still trying to find a mechanism by which HIV can be destroying far more immune system cells than it actually infects.

In a better world, the journals would be full of demands for a fundamental reconsideration of a premise adopted in such dubious circumstances, especially now that the campaign based on that premise has resulted in so much scientific failure. But too many reputations are at stake for that to be allowed to happen. Instead the U.S. president met with the AIDS lobbies in December 1995 to reassure them that the national effort will continue along the same lines, but with more money. The scandal of the circus of death is not so much that essential scientific procedures were short-circuited in a time of emergency as that after a decade of failure the researchers are unwilling to go back to the starting point to figure out where things began to go wrong.

Elinor Burkett is like many other reporters who have become emotionally involved with the tragedy of AIDS. She provides a lot of the important facts but in the end can do no more than express her frustration that things have turned out so badly. What this story needs is a reporter who can describe the scientific fiasco that resulted when activists succeeded in pressuring scientists to commit themselves to a theory and a drug before they understood the nature of the disease.

19

Genius & Plod

DO YOU WANT TO KNOW THE SECRET OF SUCCESS? SOME PEOPLE succeed on a grand scale and become world-famous. Others are just as successful but on a smaller scale. Historians and biographers do not write about their lives, but God knows what they accomplished. If you want to accomplish great things on either scale, the two qualities you need to cultivate come to the surface in Martin Gilbert's In Search of Churchill *(Wiley, 1994) and Drusilla Scott's* Everyman Revived: The Common Sense of Michael Polanyi *(Eerdmans, 1995). My essay was first published in* Books & Culture, *May/June 1997.*

During the Boer War in 1899, the twenty-five-year-old Winston Churchill made a spectacular escape from captivity. A British officer, writing to congratulate him, predicted that Churchill would someday be prime minister. "You possess the two necessary qualifications; genius and plod. Combined I believe nothing can keep them back." It's a great summary of what gave Churchill the power to survive reversals and accomplish great deeds. He had the creative instinct to know what had to be done in a tough situation and the dogged determination to do (or oversee) the detailed work required to put his grand ideas into practice.

That letter is one of many gems to be found in Martin Gilbert's *In Search of Churchill*. Gilbert is the author of the standard eight-volume biography of Churchill, and this additional memoir is the story of how he tracked down the truth about the great man. An account of how a biography was researched might be expected to make dull reading, but not when the biography is of Churchill and not when the biographer is as good as Martin Gilbert. Here is a taste of the fascinating questions Gilbert investigated: Was Churchill to blame for the Dardanelles fiasco in 1915, or was he made the scapegoat for political reasons? What were his work habits, and how much did he really drink? Who gave him the secret information he used to such devastating effect against the Baldwin and Chamberlain governments during the appeasement period? Each of these questions led Gilbert to fascinating documents and still more fascinating people.

To whet your interest, here are some answers. The Dardanelles operation was a brilliant naval concept that might have shortened the useless slaughter on the Western Front if Churchill had been supported by subordinates and cabinet colleagues as steadfast as himself. The naval operation failed because the admiral on the spot refused to go forward after an initial failure. Lord Kitchener's army then botched their part, and the temperamental First Sea Lord "Jackie" Fisher alerted the Conservative opposition to the disarray in the government by fleeing from London in a funk. Churchill lost his job as First Lord of the Admiralty because Liberal prime minister H. H. Asquith caved in to David Lloyd George's demands for a coalition government, and the Conservatives demanded Churchill's ouster as their price. Asquith, who was sixty-two years old, lost the will to fight because at the moment of crisis he was jilted by his twenty-seven-year-old heartthrob Venetia Stanley, to whom he wrote wildly indiscreet letters during cabinet meetings. Venetia's daughter gave Gilbert the letters.

Churchill did like to drink, but his work schedule amply proves that he knew when to stop. Even when he was out of office, he habitually worked after dinner until two or three in the morning, dictating journalism or historical books to relays of secretaries. In office he put in even more hours of intense concentration, bombarding subordinates with succinct memos that told them that the boss knew and cared about what they were doing. Obviously he kept a clear head. Gilbert also tells poignant stories of interviews with diplomats and military officers who risked their careers by informing Churchill of British unpreparedness, seeing as he did the disaster that lay ahead if the government and public did not wake up to the danger posed by Hitler.

That ought to be enough to make you want to read the book, so here's another hero of freedom to admire. If Churchill illustrated the power of genius and plod in statecraft, Michael Polanyi had the most to say about this combination in the not-so-different world of science. Polanyi was a distinguished chemist who turned to philosophy to correct the distortions of the materialist reductionism that has infected science in the twentieth century. His own writings are not always accessible to ordinary readers, so I am glad to see a new paperback edition of Drusilla Scott's popularization. It comes with an introduction by the distinguished theologian Lesslie Newbigin, who was heavily influenced by Polanyi.

The Hungarian-born Polanyi escaped from Central Europe just before Hitler came to power, and settled in England as professor of chemistry at Manchester University. According to Scott, Polanyi's unease with scientific materialism began with a 1935 conversation with Nikolai Bukharin, then a powerful Soviet communist theoretician and subsequently a victim of Stalin's purges. Bukharin told Polanyi that "under socialism the conception of science pursued for its own sake would disappear, for the interests of scientists would spontaneously turn to the problems of the current five-year plan." Polanyi was horrified not only by the crude instrumentalism of the Soviet attitude but by the denigration of the creative spirit that follows even in democratic societies from the tendency to see "science" as a purely objective enterprise that filters out the personal element. He turned from chemistry to philosophy to show that science as he knew it was inseparable from faith, inspiration and freedom.

The philosophy of science Polanyi criticized was largely concerned with explaining the plod of science, the testing that ensures the objectivity of scientific knowledge. Bertrand Russell and other worshipers of objectivity presented a picture in which facts were collected, hypotheses were proposed to account for the facts, and theories were preserved or discarded depending on whether they survived empirical testing. This fact-centered view of science is encapsulated by Russell's story of how Galileo supposedly dropped weights from the leaning tower of Pisa to prove that Aristotle was wrong about falling bodies. Actually Galileo never did this; like the young Einstein, he thought much and experimented little, and was not disturbed by the fact that some experiments he *had* done tended to support the opposite conclusion.

Polanyi was more concerned with the "genius" side of science, the intuitive certainty that tells a creative thinker like Galileo or Newton how things must really be and that enables them to pursue an intellectual program for years or decades despite fierce opposition and experimental discouragements. He argued that "we know more than we can say" and described a tacit knowledge that enables us to sense the outlines of reality without going through logical steps. A child learns to recognize faces and to ride a bicycle without the slightest theoretical knowledge of how such things are possible. Creative scientists likewise sense a hidden pattern behind the puzzling chaos of uninterpreted facts and go to the laboratory knowing what they mean to find.

Polanyi accordingly viewed science as akin to other kinds of human creativity rather than a separate world set apart by its commitment to objectivity. Scott illustrates this with the story of Marie Curie, who sensed with a passionate intensity the existence of an atomic property she would later name radioactivity and slaved for years under horrible conditions to prove against the opposition of conventional physicists that she was right. Marie's daughter Eve describes her mother looking at night with worshipful eyes at the first sample of radium, which glowed blue in the dark. As Scott comments, "In other times she might have been a saint or martyr, but being of her time and place, what she hungered after was scientific truth."

There is a picture of genius in that story, but there's also a lot of plod. The hard work of checking and verifying is every bit as essential as the flash of insight, because intuitive geniuses can be very wrong. Even traditionalists who argue against radical new ideas are a necessary part of the creative project, as were the Aristotelians who opposed Galileo—up to the point where they lost their faith in free inquiry and called in government power to suppress a new idea. A genius is likely to go very wrong if he or she attains the power to dispense with the plod and to put ideas into effect before they have been thoroughly checked out by critics.

I suspect we will see many examples of that principle now that it has become almost a common practice for scientists in high-profile fields to announce new discoveries at press conferences, before the scientific community has an opportunity to find the errors. The value of plod is also illustrated in the career of Winston Churchill, particularly when one compares his success to the disasters that Hitler and Stalin perpetrated with their unchecked power.

One of Churchill's personal assistants explained to Martin Gilbert why his boss was so much more effective as a war leader in World War II than in World War I. In both wars Churchill was always pressing some plan for a bold offensive thrust. "He pushed and pushed and pushed, which was all to the good . . . provided he had people to keep him on the rails. He didn't at the Admiralty [in World War I]; he dominated the Board. He did have as Prime Minister, with [General Sir Alan] Brooke and the Chiefs of Staff. That is one of the reasons why we won the war" (pp. 179-80). Of course it was Churchill himself as prime minister who was responsible for the quality of those military advisers.

That's why people like Winston Churchill and Marie Curie could accomplish what seemed to be impossible. The genius was there first, but then the plod was done right. When the two are properly combined, nothing can keep them back.

20

Facing Orthodoxy

I SUPPOSE I COULD BE ACCUSED OF A LACK OF OBJECTIVITY IN THIS review of Frederica Mathewes-Green's Facing East: A Pilgrim's Journey into the Mysteries of Orthodoxy *(HarperCollins, 1997) and* Not of This World: The Life and Teaching of Fr. Seraphim Rose *(Fr. Seraphim Rose Foundation, 1993). The fact is that I found both the Mathewes-Green family and Father Seraphim Rose so immensely appealing that I wanted to write an appreciation rather than a critique. No doubt there are difficult theological questions that Calvinists, Thomists and others will want to debate with Orthodox theologians. I'm all for vigorous debate on a proper occasion, but first I want to celebrate the treasure that we hold in common. This review first appeared in* Books & Culture, *September/October 1997.*

After fifteen years as an Episcopal priest, Gary Mathewes-Green could no longer tolerate being under the authority of apostate bishops. He and his wife Frederica, both adult converts to Christianity who had attended seminary together, began looking for a denomination that still honored the traditional creeds and moral principles. The dissident Anglican branches wouldn't do because Gary "felt he couldn't climb further out from the branch to a twig; if anything, he had to return to the trunk." The couple briefly considered the Roman Catholic Church, which allows married priests in Gary's situation, but were repelled by some of the theology, the authoritarianism and the prospect of working under the supervision of people whose thinking resembled that of the Episcopal bishops whom they were fleeing.

Gary eventually came to the Orthodox Evangelist Father Peter Gillquist, who answered his theological questions, convincing him that Orthodoxy teaches salvation by grace, not works. Frederica remained reluctant for a while to desert the sinking ship of liberalized Anglicanism, reasoning that there was a special need for chaplains on the deck of the *Titanic*. She also says that it is typical among couples converting to Orthodoxy for the husband to be gung-ho from the start and for the wife to take more time getting used to the idea. True to form, Frederica now can't imagine ever *not* being Orthodox: "I tasted and saw, and nothing can compare."

Facing East gives readers a chance to taste Frederica's experience and to compare it with their own. It recounts a year in the life of Father Gary's young missionary congregation in the Baltimore area, a family diary of a liturgical year. I found it sufficiently charming to read aloud to my wife over several weeks in our after-dinner routine. We are Presbyterians who are just as satisfied with our local church (but not our denomination!) as Frederica is with her Orthodox community. Although our ship isn't sinking, we still found much in her account to admire.

For one thing, Orthodoxy provides a magnificent *aesthetic* experience. Worshipers absorb the faith not by hearing about it but by reliving the gospel and the passion in the liturgy. This gives them a sense of contact with the historic Christian tradition that is often missing in services centered on the sermon and more closely tied to contemporary culture. Second, Orthodoxy is *demanding*. Participating in the fasts and in the long services (often standing) discourages the attitude, so prevalent among Protestants, that going to church should be something like watching television.

Finally, the Mathewes-Green parents seem to have persuaded their daughter and two sons to share a good deal of their enthusiasm. I need to hear of no further wonders. Those children are potentially more impressive answers to prayer than a thousand miraculously renewed icons.

Did I say that Orthodoxy as practiced by the Mathewes-Green family is demanding? Not if you compare it with the disciplined life of Seraphim Rose, a character straight out of the days of the desert fathers. Born Eugene Rose in San Diego in 1934, he came to San Francisco in the 1950s to seek wisdom of the gnostic kind, studying Eastern lore under Alan Watts. Eugene had the makings of a superior academic mind, including an amazing gift for learning languages. He also had a devotion to seek Truth rather than fashionable knowledge and to live for God rather than for a career. This inherent sanctity made him unsuitable for a life in the mind games of academia. In fact, it made him unsuitable for a career even in the Orthodox Church, where he was constantly in conflict with manipulative bishops.

Eugene virtually stumbled into Orthodoxy, fell under the influence of a saintly, miracle-working prelate called "Archbishop John" and saw straight through his church's flawed exterior into the patristic understanding of Christianity at its heart. He never looked back. With his friend Gleb (later Abbot Herman) he founded a monastery in the northern California mountains west of Redding, living there an arduous life of monastic asceticism and scholarship. Father Seraphim died of an intestinal infection in 1982, at the age of forty-eight, leaving volumes of inspired but loosely organized writings, mostly in the form of lecture notes or articles published in the journal *Orthodox World*. I cannot even begin to evaluate his achievement in this brief essay, except to say that I have rarely encountered so penetrating an intellect combined with so generous a spirit. His biography by a brother monk may seem overlong for some readers, but it is packed with fascinating details I wouldn't have wanted to miss.

One common criticism of Orthodoxy is that it reflects a "dark ages" mentality. Father Seraphim would have been proud to admit that he was trying to recapture the mindset of the early Christian centuries, when the church fathers and seven great councils set down the Christian faith that came to be fractured by schisms in the second millennium. I was taught to see pre-Reformation church history as the story of the Church of Rome, with Augustine and Aquinas leading to Luther and Calvin. From the Orthodox viewpoint the main story is not Rome but a turbulent, glorious millennium of church councils and inspired patristic scholars, followed by a tragic second millennium of schisms and decline. Frederica Mathewes-Green summarizes it eloquently:

> For the first thousand years, the thread of Christian unity was preserved worldwide through battering waves of heresies. The method was collegial, not authoritarian; disputes were settled in church councils, whose decisions were not valid unless "received" by the whole community. The Faith was indeed common: what was believed by all people, in all times, in all places. The degree of unity won this way was amazing. Though there was some local liturgical variation, the Church was strikingly uniform in faith and practice across vast distances, and at a time when communication was far from easy. This unity was so consistent that I could attribute it to nothing but the Holy Spirit. (*Facing East*, p. xvii)

When the unity of Christendom was broken and papal autocracy substituted for collegial deliberation, the Western Church was free to develop in a direction that led to such disasters as the Crusades, the Inquisition and the sale of indulgences. The Protestant Reformers meant to return to the roots of Christian belief, but their formula of *sola Scriptura* failed to prevent waves of further schisms. Today modernist science and postmodernist philosophy make a mockery of the Protestant fundamentals. When even many Christian institutions teach the interpretive techniques of Jacques Derrida and Michel

Foucault, against an assumed background of scientific naturalism, the words of Scripture (or any other text) can lead only to nihilistic conclusions. If "in the beginning" there existed only impersonal laws and particles, then there is no Truth. In a materialistic universe, only power and pleasure are worth pursuing.

Whatever Protestants may think of specific Orthodox practices, we should respect the motives that brought people like the Mathewes-Greens and Seraphim Rose to Orthodoxy. At bottom they are the same motives that launched the Reformation. There is a deeply felt need to dig beneath centuries of accumulated accommodation to the spirit of this world, to rediscover the treasure of authentic gospel truth that was proclaimed and defined at the beginning. Whether Orthodoxy has the right answers or not, it is profoundly attractive to people who are asking the right questions and who want to find the trunk of the tree rather than crawl farther out on a branch.

One thing we can learn from Orthodoxy is to take the long view of Christian history and see the Reformation as one episode in a much bigger story. The first Christian millennium could be called the Age of Constitution-Making. The great councils that framed the creeds and rejected the heresies were often rowdy affairs, but they achieved wonders by the Holy Spirit. Papal absolutism was not a product of this collegial process but a repudiation of it.

The second millennium was the Age of Schisms. It began with the papal legate's excommunication of the patriarch of Constantinople in 1054, reached its nadir with the sack of Constantinople and the atrocities of the wars of the Reformation, and ended with the fall of communism and (I predict) the exposure of scientific materialism as an absurdity. Throughout the twentieth century Christianity seemed doomed to wither away under the devastating critique of scientific investigation. In the end it was materialism that withered.

What name shall we give the third millennium? I like to think that we are coming to an Age of Reconstitution. Christianity is not dead or dying but poised for a new beginning in a world that needs the good

news more than ever. We need to stop multiplying schisms, set aside
the tools of worldly power and give the Holy Spirit a chance to help
us rediscover the truth that once united us. Those of us who are not
inclined to join the converts to Orthodoxy can nonetheless rejoice to
have them as worthy partners in that great work of healing.

21

The Law Written
on the Heart

PEOPLE WHO SAY THEY DON'T BELIEVE IN NATURAL LAW JUST DON'T understand what they are saying. Everybody believes in some fundamental principles that are independent of the prejudices that derive from particular cultural settings. Even the relativist principle that there is no universal moral truth leads to ethical conclusions, such as that it is wrong for you to impose your moral principle on me because of your mistaken belief that those principles are universally valid. The interesting question is not whether natural law (moral principles accessible to reason) exists but on what starting point it should be based. In Written on the Heart *(InterVarsity Press, 1997; my review was first published in* Books & Culture, *July/August 1997) J. Budziszewski addresses the main issue more directly than any other contemporary thinker.*

Paul wrote in Romans 2:15 that Gentiles who know nothing of Moses or Christ may nonetheless show by their deeds "that the requirements of the law are written on their hearts, their consciences also bearing witness, and their thoughts now accusing, now even defending them." J. Budziszewski explains that this law is what philosophers call the "natural law." It is the bedrock moral understanding that we can't *not* know, however hard we try to evade that knowledge, because our consciences bear witness to it.

When our consciences accuse us and we are unwilling to repent, all we can do is to smother our knowledge with rationalizations and recruit others to vice. As Paul said in Romans 1:32, "Although [depraved people] know God's righteous decree that those who do such things deserve death, they not only continue to do these very things but also approve of those who practice them." Just as misery loves company, sin craves social approval.

Most of *Written on the Heart* consists of a highly readable and stimulating survey of the history of natural-law thinking from Aristotle to the present. Aristotle, who knew nothing of the Judeo-Christian God, developed his commonsense ethical philosophy by examining the actual practices of people who were reputed to be wise and happy. Thomas Aquinas melded Aristotelian ethics with Roman legal scholarship and Catholic doctrine to create a synthesis that still has a powerful attraction for those who study it sufficiently to master Thomas's categories. John Locke, who meant to find a stronger basis for law, unintentionally undermined the project by grounding knowledge on sense experience exclusively. Eventually his empiricism led to utilitarianism, which attempted to rebuild moral philosophy on a dismally inadequate foundation, namely our desire for pleasure and aversion to pain. Utilitarianism led inexorably to pragmatism and relativism, in reaction to which some Roman Catholic natural law philosophers (especially Germain Grisez and John Finnis) have attempted to revive natural law theory on a secular basis.

I would have added another major figure to this historical survey: the immensely influential American jurist Oliver Wendell Holmes Jr. Justice Holmes thought of law as a science that, like the natural sciences, rigorously excluded such irrelevancies as morality and metaphysics. "If you want to know the law and nothing else," he wrote in

The Path of the Law (1897), "you must look at it as a bad man, who cares only for the material consequences which such knowledge enables him to predict, not as a good one, who finds his reasons for conduct . . . in the vaguer sanctions of conscience."

From the standpoint of the practicing lawyer, Holmesian law is nothing more than "the prophecies of what the courts will do in fact." The bad person and the good person alike will guide their conduct by such prophecies, because neither wishes to come into conflict with the organized force of the state. From the standpoint of the rational judge or legislator, the rules that the prudent bad person observes are themselves derived from policy sciences like economics and psychology. The ultimate purpose of law is to achieve whatever goals (such as prosperity and safety) the public sees fit to endorse through the political process. Whether people's inclinations are good or bad is of little concern, because bad and good alike can be made to obey the law.

Holmes downgraded tradition as a source for law, famously remarking, "It is revolting to have no better reason for a rule of law than that it was so laid down in the time of Henry IV." He had an even lower regard for notions of morality, considering them unscientific relics of outdated religious traditions. He conceded that a law which too blatantly transgresses a community's moral standards might be unenforceable, but then trivialized the point by illustrating it with the comment "I once heard the late Professor Agassiz say that a German population would rise if you added two cents to the price of a glass of beer." Never one to pull his punches, Holmes mused, "I often doubt whether it would not be a gain if every word of moral significance could be banished from the law altogether, and other words adopted which should convey legal ideas uncolored by anything outside the law." In place of the moral law written on the heart, Holmes gave us state coercion based on science. That's about as far away from Thomas Aquinas as you can get.

Such nihilism also puts the political community and the pirate gang on the same moral footing, since both are based on pragmatism and coercion. It is no wonder that people continue to search for a new kind of natural law that agnostic modernists can accept, but the immediate prospects are not promising. Germain Grisez postulates that there are seven basic forms of the good, including justice, friendship, holiness, life (including procreation), knowledge and skill. Each of these is said to be "irreducible," so that it is forbidden, for example, to sacrifice one basic good to achieve another. Budziszewski's brief critique of this complex theory is devastating: its rules are arbitrary, and they threaten to turn the easy questions (like whether it is permissible to adopt a celibate lifestyle in order to pursue holiness or knowledge) into unresolvable dilemmas.

James Q. Wilson's 1993 book *The Moral Sense* asserts that "we have a moral sense [and] most people rely on it even if intellectuals deny it, but it is not always and in every aspect of life strong enough to withstand a pervasive and sustained attack." That may sound like the law written on the heart, which can be obscured but never erased. As a scientific materialist, however, Wilson cannot ground the moral sense in anything more solid than "feelings," meaning emotional reactions which he deems to have been created by natural selection. The well-known problem with this approach is that we have many conflicting feelings, some of which (like avarice and lust) hardly qualify as moral. Wilson has to distinguish the truly moral feelings from their "wilder rivals," much as the utilitarian John Stuart Mill tried to distinguish the higher pleasures from the lower ones. These moves invoke an objective moral law by which feelings or pleasures can be evaluated, but where is such a law to be found?

The best part of *Written on the Heart* is chapter thirteen, where Budziszewski provides a brilliant "Christian appraisal of natural law theory." Natural law is not in any sense a substitute for divine revelation or saving grace. For a Christian the Bible is the paramount authority on moral questions, but the Bible itself teaches that God has a witness (general revelation) to the pagans. Indeed, the heartfelt admission that there is a moral law and that we have violated it is often the first step that brings the unbeliever to faith. C. S. Lewis's apologetic in *Mere Christianity* takes exactly this approach. Of course the law written on the heart is obscured by what psychologists call "denial," and modernists far surpass the ancient pagans in inventing strategies for denial. In Budziszewski's words: "With a head filled with false sophistication that tells him that right and wrong are invented by culture and different everywhere, the new sort of pagan mistrusts his own conscience and views guilt as a sign of maladjustment that therapy will remove."

Most modern ethical thinking, Budziszewski explains, goes about matters backwards. Modernists assume that the problem of sin is mainly *cognitive*—that we don't know the moral law and are doing our best to find it out. Unfortunately for us, the problem is mainly *volitional*. We know well enough the difference between right and wrong, but we obscure our understanding so we can do as we please. That is why the primary task of Christian natural law philosophy is not to prove the existence of the moral law but to expose the devices of the heart by which we conceal the truth from ourselves.

The concept of natural law makes sense only if our lives have a purpose. Consider two influential statements of the human condition. The first comes from the neo-Darwinist George Gaylord Simpson: "Man is the result of a purposeless and natural process that did not have him in mind." This is modernity's official doctrine of creation, and it provides no foundation on which moral reasoning can build. As accidental byproducts, we might as well do whatever gratifies our strongest feelings or helps us to get whatever it is that we happen to want. All else is pious humbug.

Now consider the famous words of the Westminster Catechism: "Man's chief and highest end is to glorify God, and fully to enjoy him forever." From that statement we know that a moral law exists, and it consists of those precepts that teach us how to achieve our chief and highest end. If we start there, we can read what is written on our hearts.

22

Making Law Sane

*THE EMINENT SOCIAL SCIENTIST JAMES Q. WILSON REMINDS ME OF MY
Berkeley colleague John Searle, who is equally eminent as a philosopher (see essay 6, "Daniel Dennett's Dangerous Idea," in this book,
and also chapter six of my* Reason in the Balance *[InterVarsity Press,
1996]). Both Wilson and Searle come the right conclusions about
many important things, but they come to them by convoluted reasoning
after starting from the wrong premises. Here I focus on Wilson's* Moral
Judgment: Does the Abuse Excuse Threaten Our Legal System?
(BasicBooks, 1997; my review was first published in Books & Culture,
November/December 1997).

During the last two centuries the dominant intellectual culture
has discarded a theistic understanding of human nature and
replaced it with a scientific—or naturalistic—understanding.
Once upon a time we thought humans were moral agents created in
the image of a supernatural God, with a divine gift of freedom and a
knowledge of God's moral order written on our hearts. Now we think
otherwise, if we obediently follow our cultural leaders. Twentieth-century science teaches us that our thoughts and actions are products of

our genetic endowment and our cultural environment. Scientific thinkers may disagree about the relative importance of biology and culture, but they agree that who we are and what we do is explained by some combination of nature and nurture. What else is there?

We might have expected such an intellectual revolution to affect the criminal law, and it has done so. Earlier ages felt comfortable with the idea that a murderer deserves to die because he has deliberately taken the life of another to further his own selfish ends. Modernists think otherwise. The murderer acted as he did because of some genetic predisposition or because he was abused as a child or otherwise mistreated by society. For modernists crime is a social problem like disease, which a rational society seeks to mitigate by identifying and treating the "root causes."

James Q. Wilson, the well-known political scientist whose previous books include *Thinking About Crime* and *The Moral Sense,* is himself a thoroughgoing modernist who nonetheless rejects the fruits of modernism in criminal law. He attacks expansions of the insanity defense and other psychological defenses like the "battered wife syndrome" that have allowed killers to escape punishment because "they couldn't help it." Beware of social scientists bearing syndromes, he warns; they are usually peddling junk science.

Although many of Wilson's criticisms of specific doctrines are sensible, the picture he paints of the overall state of criminal law is grotesquely misleading. The "abuse excuse" is not out of control, the insanity defense is not expanding, and the law is not losing its moral compass to wallow in pseudoscientific explanations of criminal behavior. Far from it. The expanded mental illness defenses Wilson criticizes had their heyday in a few states a generation ago, and they have been repudiated whenever they had tangible consequences. Criminal sentences are much longer than they used to be, and prison is frankly regarded as a punitive remedy. Even the death penalty, which seemed on the way out thirty years ago, has made a comeback. Relatives of murder victims now have standing to appear in sentencing and parole hearings, and their arguments for retributive justice are heeded.

That Wilson is describing anomalies rather than the norm is illustrated by one of his prime examples, the infamous Menendez brothers case. The brothers, who cold-bloodedly murdered their wealthy parents, claimed that they acted in self-defense because the parents were planning to kill them to cover up a history of sexual abuse. The defense managed to muddy the waters enough with this preposterous story that the jurors could not agree on a verdict. That was embarrassing for the criminal justice system, but it didn't do the defendants a bit of good. On retrial the judge ran a much tighter ship, and the jury had no difficulty convicting both brothers of first-degree murder. The second trial, not the first one, is typical of what happens to cold-blooded murderers. Wilson admits this but says that the exceptional, high-profile cases disproportionately affect public perceptions of the system. True enough, but the whole thrust of Wilson's argument furthers the misperception that abuse excuses are expanding out of control.

Wilson even distorts facts to support the false picture of a lenient system that allows every conceivable excuse. He writes, "Sentences for homicide are . . . relatively short (in the 1990s around six years for the nation and four years in California)." Tell that to the Menendez brothers, who (like most aggravated murderers in California) are serving life without eligibility for parole. Wilson seems to have confused the average prison terms of persons released on parole for manslaughter (i.e., the least culpable killers) with the typical prison term for murderers, who serve many years before parole if they ever get paroled. If you are thinking of committing a murder in California, it would be better not to rely on legal advice from Professor Wilson.

A criminal justice system that tries thousands of cases every week is bound to include examples of juries who refuse to convict because they dislike the victim or feel sorry for the defendant or swallow tall tales about police conspiracies. I agree that we should not encourage this sort of thing by exposing jurors to far-fetched psychological theories, but lawmakers have largely learned this lesson. Consider the recent history of the insanity defense, for example. Anglo-American law traditionally allows only a very narrow insanity defense, requiring defendants to prove that due to mental illness they were incapable of understanding the wrongfulness of their act (usually homicide). A major goal of modernist law reformers was to broaden the defense to put the burden of proof on the prosecution and to allow acquittal of defendants who could not control their conduct. Since modernist science assumes that all behavior is "caused" by some combination of nature and nurture and banishes the freely choosing moral subject to the realm of metaphysics, the "lack of free will" defense was potentially open-ended.

The expanded insanity defense was endorsed by leading experts and enacted in the federal system and many states. (In California, judges left the old insanity rule unchanged but introduced the new philosophy directly into the law of murder by saying that a defendant lacked "malice aforethought" if he couldn't control his conduct.) The new rules lasted just until they succeeded in generating outcomes the public recognized as crazy, including the insanity acquittal of John Hinckley—who shot President Reagan and his press secretary in hopes of attracting the notice of the movie star Jodie Foster. Of course Hinckley's motivation really was loony, but he also knew the wrongfulness of what he was doing and chose to do it. Public opinion promptly forced a change back to the old rules, with additional measures designed to ensure that defendants acquitted for insanity would be confined just as securely as if they had been convicted. John Hinckley is still behind bars, and going nowhere.

What is particularly fascinating about the traditional insanity doctrine, called the M'Naghten Rule by lawyers, is that it is based straightforwardly on assumptions derived from biblical theism. Humans are seen as endowed with an innate understanding of the difference between *moral* right and wrong—meaning an absolute moral standard that is independent of legal rules. The law holds us responsible if we choose wrong instead of right, just as God does—and science does not. Criminal defendants are excused for insanity only if this innate capacity for moral understanding is so damaged that they are comparable to small children, who do not grasp what killing means even if they pick up a loaded pistol, point it at a playmate and pull the trigger. (A California six-year-old was recently found incapable of committing attempted murder after he beat a baby almost to death. No one protested the decision.) Insanity in this restricted sense saves a killer from the death penalty, but it does not lead to freedom, because an adult who does not know right from wrong belongs in custody.

Whatever scientific naturalists may say, criminal law has found it necessary to assume that humans are moral agents created in the image of God, with a divine gift of freedom and a knowledge of God's moral order written on our hearts. Even James Q. Wilson, who doesn't believe the premise, likes the conclusions that follow from that premise. When you are dealing with human beings, naturalism is a bust—especially as a methodology.